Euro-TMCS I: Theory, Modelling and Computational Methods for Semiconductors

Journal of Physics: Conference Series Volume 609

Granada, Spain
28-30 January 2015

ISBN: 978-1-5108-0752-5
ISSN: 1742-6588

Printed from e-media with permission by:

Curran Associates, Inc.
57 Morehouse Lane
Red Hook, NY 12571

Some format issues inherent in the e-media version may also appear in this print version.

Printed by Curran Associates, Inc. (2015)

For permission requests, please contact the Institute of Physics
at the address below.

Institute of Physics
Dirac House, Temple Back
Bristol BS1 6BE UK

Phone: 44 1 17 929 7481
Fax: 44 1 17 920 0979

techtracking@iop.org

Additional copies of this publication are available from:

Curran Associates, Inc.
57 Morehouse Lane
Red Hook, NY 12571 USA
Phone: 845-758-0400
Fax: 845-758-2634
Email: curran@proceedings.com
Web: www.proceedings.com

TABLE OF CONTENTS

Journal of Physics: Conference Series **609** (2015) 011001 doi:10.1088/1742-6596/609/1/011001

Introduction to EURO-TMCS I

The present issue contains a selection of the best contributed works presented at the first Euro-TMCS conference (Theory, Modelling and Computational Methods for Semiconductors, European Session). The conference was held at Faculty of Sciences, Universidad de Granada, Spain on 28st-30st January 2015.

This conference is the first European edition of the TMCS conference series which started in 2008 at the University of Manchester and has always been held in the United Kingdom. Four previous conferences have been previously carried out (Manchester 2008, York 2010, Leeds 2012 and Salford 2014).

Euro-TMCS is run for three days; the first one devoted to giving invited tutorials, aimed particularly at students, on recent development of theoretical methods. On this occasion the session was focused on the presentation of widely-used computational methods for the modelling of physical processes in semiconductor materials. Freely available simulation software (SIESTA, Quantum Espresso and Yambo) as well as commercial software (TiberCad and MedeA) were presented in the conference by members of their development team, offering to the audience an overview of their capabilities for research.

The second part of the conference showcased prestigious invited and contributed oral presentations, alongside poster sessions, in which direct discussion with authors was promoted. The scope of this conference embraces modelling, theory and the use of sophisticated computational tools in semiconductor science and technology. Theoretical approaches represented in this meeting included: Density Functional Theory, Semi-empirical Electronic Structure Methods, Multi-scale Approaches, Modelling of PV devices, Electron Transport, and Graphene. Topics included, but were not limited to: Optical Properties of Quantum Nanostructures including Colloids and Nanotubes, Plasmonics, Magnetic Semiconductors, Photonic Structures, and Electronic Devices.

The Editors

Acknowledgments: We would like to thank all participants for making this a very successful meeting and for their contribution to the conference programme and these proceedings. We would also like to acknowledge the financial support from Universidad de Granada, the CECAM UK-Hartree Node, project TEC2013-47283-R of Ministerio de Economía y Competitividad, and the company Materials Design (distributors of MedeA Software).

Conference Organising Committee:
Francisco M. Gómez-Campos (Co-chair, Universidad de Granada)
Salvador Rodríguez-Bolívar (Co-chair, Universidad de Granada)
Stanko Tomić (Co-chair, University of Salford)

Journal of Physics: Conference Series **609** (2015) 011002 doi:10.1088/1742-6596/609/1/011002

Peer review statement

All papers published in this volume of *Journal of Physics: Conference Series* have been peer reviewed through processes administered by the proceedings Editors. Reviews were conducted by expert referees to the professional and scientific standards expected of a proceedings journal published by IOP Publishing.

Journal of Physics: Conference Series **609** (2015) 012001 doi:10.1088/1742-6596/609/1/012001

Light induced structural changes in $CH_3NH_3PbI_3$ Perovskite Solar Cells

Luca Bertoluzzi[1]
Photovoltaics and Optoelectronic Devices Group, Physics Department, Jaume I University, 12071 Castelló, Spain.

E-mail: bertoluz@uji.es

Abstract. Perovskite solar cells have recently become one of the most promising alternative to the current solar cell technologies by achieving efficiencies superior to 20%. Owing to their distortable structure these devices have proven to show unusually slow relaxation processes upon illumination. We have identified this phenomenon in the frequency domain by impedance spectroscopy (IS) and in the time domain by transient photovoltage decay (TPD) and open circuit voltage decay (OCVD). However the origin of this process is still unclear although three main hypotheses have been proposed in the literature: i) high internal electric field, ii) ferroelectric domains and iii) ion/defect diffusion. In this conference paper we first present our works on light induced structural changes in perovskite and then discuss the main hypotheses previously mentioned with the insight of first principle calculations.

1. Introduction

$CH_3NH_3PbI_3$ perovskite has proven to be very successful as a light absorber, leading to solar cells with efficiencies as high as 20.1% [1]. This material is composed of an organic methyl ammonium cation, $CH_3NH_3^+$, which is located in an inorganic framework composed of four PbI_6 octahedra. $CH_3NH_3^+$ has been shown to be bound to the inorganic PbI_6 octahedra through Van der Waals interactions. In such material the lowest conduction band is formed of the Pb^{2+} *6p* orbitals, while the highest valence band is constituted by the I^- *5p* orbitals [2]. Moreover, this perovskite structure displays a gap of 1.5 eV. Note that the gap of lead halide perovskite is strongly influenced by the nature of the halide anion [3, 4].

Structurally speaking this type of perovskite undergoes three phase transitions with temperature. $CH_3NH_3PbI_3$ is in the orthorhombic phase at low temperature ($T < 161.4$ K), in the tetragonal phase at higher temperatures below 330.4 K and in the cubic phase above 330.4 K [2]. At room temperature $CH_3NH_3^+$ is therefore in the tetragonal phase and its crystallographic structure is distortable. This latter point is of considerable interest for photovoltaic applications since various studies have shown that perovskite undergoes structural changes under illumination.

In the following we first summarize the works published in our group, showing through electrochemical techniques the light induced structural changes in the perovskite absorber. Then we discuss, with the insight of first principle calculations, the main possible causes that can be found in the literature for such peculiar behaviour. Finally we give our conclusions on the topic.

[1] To whom any correspondence should be addressed.

Journal of Physics: Conference Series **609** (2015) 012001 doi:10.1088/1742-6596/609/1/012001

2. Light induced structural changes in $CH_3NH_3PbI_3$ perovskite.

Recently we have probed the relaxation processes in $CH_3NH_3PbI_3$ solar cells by means of electrochemical techniques. Mainly we have used impedance spectroscopy (IS), transient photovoltage decay (TPD) and open circuit voltage decay (OCVD). Before entering into the details of our results, let us briefly explain the basic principles underlying these techniques.

2.1. Probing $CH_3NH_3PbI_3$ structural changes by electrochemical techniques

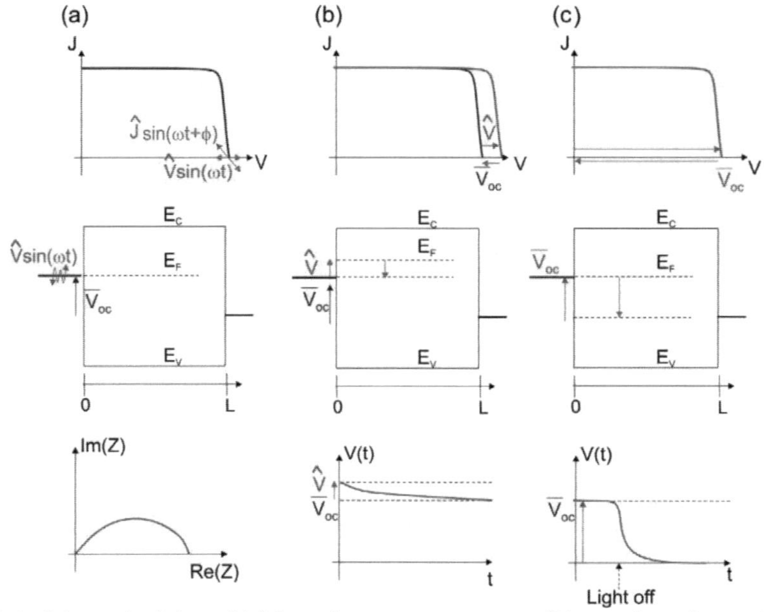

Figure 1. Underlying principles of (a) impedance spectroscopy, (b) transient photovoltage decay and (c) open circuit voltage decay. For each method, we show how the J-V curve (first row) and the energy diagram (second row) are modified when applying these techniques and we show how the response signal is typically represented (third row). Note that for simplification of the energy diagram, only the Fermi level of the minority carrier is represented. In this scheme blue lines and arrows refer to the steady state, red ones to the state resulting from the excitation (by light or voltage) and green ones to the response of the system after excitation.

In figure 1, we present a scheme that summarizes the techniques mentioned above. The strength of these techniques relies in their ability to decouple the different faradic (i.e. resistive) and capacitive processes that occur in a solar cell, which is impossible to achieve through steady state measurements. IS and TPD are small perturbation methods unlike OCVD. The main difference between IS and TPD relies in the nature of the small perturbation. In IS the perturbation is a sinusoidal, frequency dependent voltage, $\hat{V}$ that is added to a steady state voltage, $\bar{V}$, and the resulting perturbation of the current, $\hat{J}$, is measured. IS measurements are usually represented in terms of impedance ($Z = \hat{V}/\hat{J}$) in the complex plane and are interpreted with equivalent electrical circuits. In TPD a small pulse or step of voltage is applied at open circuit and one monitors the open circuit voltage decay to the steady state voltage. These measurements are interpreted in terms of decay times, which can be related to the electrical elements used for the interpretation of IS. Finally, the OCVD technique differs from the TPD in that the large perturbation of voltage decays to the ground state and not to a given steady state. The advantage of this technique over the TPD is that the measured signal is less noisy. It should be noted however that the OCVD is not necessarily reconstituted from the superposition of TPDs and both techniques may give different information.

Journal of Physics: Conference Series **609** (2015) 012001 doi:10.1088/1742-6596/609/1/012001

2.2. Detection of anomalous relaxation processes in $CH_3NH_3PbI_3$ upon illumination.
Our impedance measurements revealed the presence of a very high capacitive effect at low frequency upon illumination of the perovskite solar cell. From this measurements we could evaluate the static dielectric constant to take values as high as $10^6 - 10^7$ whereas in dark conditions it was assessed to be around $10^2 - 10^3$ [5]. These measurements suggest that light activates a new polarization process, which is coupled to the classical electronic drift diffusion recombination [6]. These measurements where confirmed by TPD where slow decay times (around 1.5 seconds) have been measured [7]. More evidence has been suggested by our recent work on the OCVD technique applied to perovskite solar cells where decay times of the order of $10 - 100$ seconds have been measured [8]. Moreover we have analyzed the shape of these decays and found out that they follow a power law time dependence, which suggests cooperative polarization/depolarization processes. Our results do shine light on a new polarization process in $CH_3NH_3PbI_3$ perovskite solar cells but the nature of this process is still unknown. For this latter purpose, first principle calculations are needed.

3. Insight from first principle calculations.
Three main hypotheses have been suggested in order to explain the high polarization of the perovskite at low frequency: the existence of a strong internal electric field, ferroelectric domains and ion/defect diffusion in the perovskite absorber. Let us quickly review and discuss these hypotheses with the insight of first principle calculations, which have been performed by other groups.

3.1. High internal electric fields.
DFT calculations have shown that effective masses for both *n* and *p* type $CH_3NH_3PbX_3$ perovskites are unusually low (0.12 – 0.14) [9, 10]. These results suggest a low density of states for electrons and holes. In fact, Manser et al. [11] have measured a density of states for the lower edge of the conduction band of the order of 10^{17} cm^{-3}, which is two orders of magnitude less than the one of silicon. It is therefore likely that strong electric fields are present in lead halide perovskite cells. Note however that these fields are strongly dependent on the contacting materials and fabrication method of the perovskite absorber. It should also be remarked that neither the theoretical nor the experimental studies explain how the generation of free carriers by illumination of the lead iodide perovskite influences this electric field.

3.2. Ferroelectric domains.
Several groups have studied the possible existence of ferroelectric domains in $CH_3NH_3PbI_3$ perovskite based solar cells. Three causes have been identified: distortion of the inorganic octahedra, off centering of the Pb^{2+} cation and rotation of the $CH_3NH_3^+$ ion in its inorganic "cage" [12−14]. However this last process is unlikely to cause the low frequency polarization observed experimentally since at room temperature the rotation time of the organic cation is of the order of picoseconds. It is worthwhile noticing that the effect of light on ferroelectric domain formation has not been considered in these studies.

3.3. Ion/defect diffusion
The possibility and conditions for a possible ion/defect diffusion in perovskite solar cells have still not been proven theoretically. However the existence of low formation energy defects has been demonstrated through DFT calculations. Wan-Jian Yin and co-workers have shown that deep defects resulting from the covalent bonds formed by Pb (dimers) and I (trimers) have a high formation energy and are consequently unlikely to be formed [15]. On the contrary, shallow defects arising from ionic interaction between $CH_3NH_3^+$ and the inorganic PbI_6 cage have a low formation energy. In particular lead vacancies and methyl ammonium interstitials are more likely to be formed. In addition shallow defects of the Frenkel type (interstitials) could act as dopants. However the one of the Schottky type (vacancies) are more likely to dominate the defect

Journal of Physics: Conference Series **609** (2015) 012001 doi:10.1088/1742-6596/609/1/012001

population with a formation energy of 0.14 eV and could reach concentrations higher than 0.4 % [16].

4. Conclusions.

Through three electrochemical techniques (IS, TPD and OCVD), we have shown that perovskite based solar cells display unusually slow relaxation processes, which are associated to the polarization of the perovskite absorber at low frequency. These findings have considerable importance for the field of photovoltaics since solar cells work in steady state condition (zero frequency). However the origin of such process is still under discussion. In the literature, three main possible reasons have been suggested: high internal electric field, ferroelectric domains and ion diffusion. The first one could be justified by the low density of states measured experimentally and calculated with DFT. The second hypothesis has been analysed by DFT and the distortion of the PbI_6 cage, the off-centering of the Pb^{2+} cation and the rotation of the $CH_3NH_3^+$ cation have been regarded as possible reasons for the creation of ferroelectric domains. Finally, ion/defect diffusion has not been demonstrated theoretically but the likely existence of shallow defects has been shown due to their low energy formation. Nevertheless none of these studies has been performed under illumination conditions, i.e. only the ground state perovskite absorber has been studied. The important problem of the structural changes induced by light therefore remains. In other words, one fundamental question is: how does light induce ferroelectric domains, ion diffusion or any other process that could cause the low frequency polarization?

Acknowledgments
The author wishes to thank Dr. Rafael S. Sanchez and Prof. Juan Bisquert for reading the manuscript and acknowledge funding from the European Union Seventh Framework Program [FP7/2007–2013] under grant agreement 316494.

[1] http://www.nrel.gov/ncpv/images/efficiency_chart.jpg.
[2] Wang Y, Gould T, Dobson J F, Zhang H, Yang H, Yao X and Zhao H 2014 *Phys. Chem. Chem. Phys.* **16** 1424-9
[3] Quarti C, Grancini G, Mosconi E, Bruno P, Ball J M, Lee M M, Snaith H J, Petrozza A and Angelis F D 2013 *J. Phys. Chem. Lett.* **5** 279-84
[4] Umari P, Mosconi E and De Angelis F 2014 *Sci. Rep.* **4**
[5] Juarez-Perez E J, Sanchez R S, Badia L, Garcia-Belmonte G, Kang Y S, Mora-Sero I and Bisquert J 2014 *J. Phys. Chem. Lett.* **5** 2390-4
[6] Bisquert J, Bertoluzzi L, Mora-Sero I and Garcia-Belmonte G 2014 *J. Phys. Chem. C* ***118*** 18983-91
[7] Sanchez R S, Gonzalez-Pedro V, Lee J-W, Park N-G, Kang Y S, Mora-Sero I and Bisquert J 2014 *J. Phys. Chem. Lett.* **5** 2357-63
[8] Bertoluzzi L, Sanchez R S, Liu L, Lee J-W, Mas-Marza E, Han H, Park N-G, Mora-Sero I and Bisquert J 2015 *Energy Environ. Sci.* DOI: 10.1039/c4ee03171g.
[9] Even J, Pedesseau L, Jancu J-M and Katan C 2014 *Phys. Status Solidi (RRL)* **8** 31-5
[10] Filippetti A and Mattoni A 2014 *Phys. Rev. B* **89** 125203
[11] Manser J S and Kamat P V 2014 *Nat. Photon.* **8** 737-43
[12] Benedek N A and Fennie C J 2013 *J. Phys. Chem. C* **117** 13339-49
[13] Brivio F, Walker A B and Walsh A 2013 *APL Mat.* **1** 042111
[14] Frost J M, Butler K T, Brivio F, Hendon C H, van Schilfgaarde M and Walsh A 2014 *Nano Lett.* **14** 2584-90
[15] Yin W-J, Shi T and Yan Y 2014 *Appl. Phys. Lett.* **104** 063903
[16] Walsh A, Scanlon D O, Chen S, Gong X G and Wei S-H 2015 *Angew. Chem. Int. Ed.* **54** 1791-4

Journal of Physics: Conference Series **609** (2015) 012002 doi:10.1088/1742-6596/609/1/012002

Electronic transport in QD based structures: from basic parameters to opto-electronic device simulations

S. Illera, J. D. Prades, A. Cirera and A. Cornet

MIND/IN^{2}UB Departament d'Electrònica, Universitat de Barcelona, C/Martí i Franquès 1, E-08028 Barcelona, Spain

E-mail: sillera@el.ub.edu

Abstract. We present a theoretical model that explains the optoelectronic response of nano-devices based on large quantum dot (QD) arrays. The model is grounded on rate equations in the self-consistent field regime and it accurately describes the most important part of the system: the tunnel junctions. We demonstrate that the ratio between the optical terms and the transport rates determines the final device response. Furthermore, we showed that to obtain a net photocurrent the QD has to be asymmetrically coupled to the leads.

1. Introduction

Silicon quantum dots (Si QD), embedded in insulator matrices such as SiO_2, have opened a new branch of possibilities in the electronics due to the novel electronic properties based on the quantum confinement. Discrete energy states appear inside the wide band gap of the insulator matrix, making possible tunable band gap devices. Concerning opto-electronic devices, the possibility to modify the energy band gap as a function of the QD radius can be exploited in order to build light absorbers for photovoltaic applications.

In order to generate a net photocurrent, the photo-generated electron has to be extracted before recombining. Thus, the charge transport from the location of photo-generation to the leads becomes a crucial point that governs the response of the system. These transport mechanisms depend on the tunneling processes being strongly dependent on the insulator material and also on the geometrical arrangement of the QDs.

2. Theoretical model

A compact model to explain ballistic electrical transport in the self-consistent field regime in these kind of systems was reported by the authors previously [1, 2]. The transport formalism was compared to Non Equilibrium Green's Function Formalism (NEGFF) obtaining similar results [3] and it also showed the possibility to combine it with Density Functional Theory (DFT) [4]. The general framework describes the electrical transport between N QDs embedded in a dielectric matrix coupled to two electrodes (L lead and R lead) as a network of multi-tunnel-junctions under external bias voltage polarization. It is based on the Transfer Hamiltonian formalism [5] to describe the currents between the different parts of the system. Assuming that the QDs are independent, the non-equilibrium distribution functions can be obtained solving a

Journal of Physics: Conference Series **609** (2015) 012002 doi:10.1088/1742-6596/609/1/012002

set of non-coherent rate equations [6]. Moreover, the effect of the trapped charge in the QDs has been included solving the transport equations and the Poisson equation self-consistently, within the self-consistent field regime, explaining a first order the Coulomb blockade effects [7]. From the previous developed formalism, large systems based on QD arrangements can be simulated in a great detail taking into account the particular properties of the QDs.
In the spirit of the rate equation type model, we assume that the optical and transport processes are independent [8]. Thus, the rate equations have been modified in order to reflect the stimulated generation/recombination carrier processes. For each energy level, they read as

$$\frac{dn_j^i}{dt} = \underbrace{\frac{2\pi q}{\hbar}|T_{Lj}|^2\rho_L\rho_j^i(f_{Lj}-n_j^i)}_{\text{Left lead contribution}} + \underbrace{\frac{2\pi q}{\hbar}|T_{Rj}|^2\rho_R\rho_j^i(f_{Rj}-n_j^i)}_{\text{Right lead contributions}}$$

$$+\underbrace{\sum_{k,i'\neq i}\frac{2\pi q}{\hbar}|T_{jk}^{ii'}|^2\rho_j^i\rho_k^{i'}(n_k^{i'}-n_j^i)}_{\text{Neighboring QDs contribution}} + \underbrace{\sum_k qR_{kj}\frac{\rho_k^i}{\rho_j^i}n_k^i(1-n_j^i) - \sum_k qR_{jk}n_j^i(1-n_k^i)}_{\text{Light terms}}, \quad (1)$$

where the superscript i and i' refer to the ith and i'^{th} QDs, respectively. j and k refer to the jth and k^{th} energy level of the corresponding QD. n_j^i is the non-equilibrium distribution function of the jth level in the ith QD. ρ_L and ρ_R are the density of states (DOS) of the leads evaluated at the energy of the energy level j, and ρ_j^i is the degeneracy of the jth energy level of the ith QD. The QD DOS includes the local potential in each QD as the solution of the Poisson equation (see Ref. [1]). f_{Lj} and f_{Rj} are the distribution functions of the leads (left and right, respectively) described by the Fermi Dirac distribution function taking into account that the external bias voltage (V) modifies the electrochemical potentials $\mu_L - \mu_R = -qV$.

$|T_{Lj}|$ and $|T_{Rj}|$ are the transmission coefficients between the QD and the leads whereas, $|T_{jk}^{ii'}|$ is the transmission coefficient among the ith and i'^{th} QDs. For the description of the tunneling processes through the oxide, we have used the WKB approximation [9] as we have described previously in Ref. [2].

The QD is treated as a finite spherical potential well within the effective mass approximation [10]. The height of the potential well is the difference between the conduction/valence band energy level of the dielectric matrix and the ones that form the QD. The width of the well is $2R$, where R is the QD radius. The carriers effective masses are assumed to have different values in the QD and in the dielectric matrix.

Concerning the optical transition rates R_{jk}, they were calculated using the Fermi's Golden Rule in the dipole approximation within the strong confinement regime [11, 12]. This is the usual treatment of the light interaction and a formal derivation can be found in many text books [13, 14].

Solving Eq. 1 in the steady state condition, the QD non-equilibrium distribution function for each energy can be obtained. Therefore, the charge stored in the ith QD is $N_i = \sum_j 2\rho_j^i n_j^i$, where we take into account the spin factor 2. Once the charge has been calculated, the Poisson equation is solved and the self-consistent solution of the local potential and the charge is imposed [1]. From the non-equilibrium distribution functions, the current through the device can be obtained as $I = \sum_i^N \sum_j \frac{2q^2}{\hbar}|T_{Lj}^i|\rho_L\rho_j^i(f_{Lj}-n_j^i)$ where the current sums over all the N QDs (i) and the corresponding energy levels of each QD (j).

3. Si QDs optical simulations: the role of the QD arrangement

Here, we are going to focus on the specific case of silicon QDs (Si QDs) embedded in a SiO_2 matrix (Si/SiO_2 QDs). The inputs needed to describe this system are: the carrier effective masses, the confinement potentials, the Si bulk band gap and the dielectric constants. The value of the parameters can be found elsewhere [2].

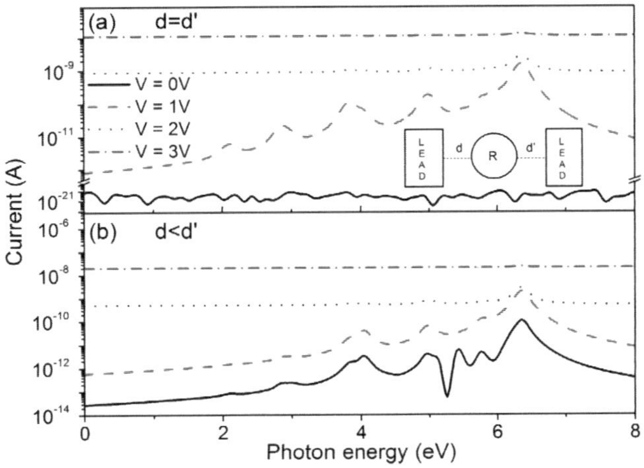

Figure 1. Photocurrent as a function of the energy of the incident light with an external applied bias voltage for (a), the symmetric case ($d = d' = 1.78nm$) and (b), asymmetric case ($d = 1.47nm$ and $d' = 2.09nm$). The scheme of the single Qd of $R = 1.05nm$ placed between the two electrodes is shown in the inset.

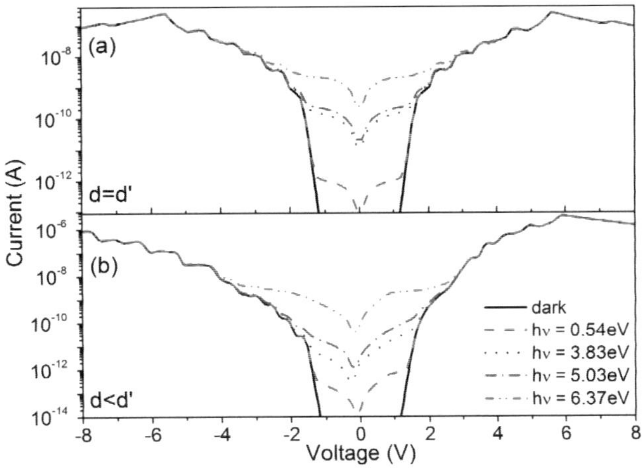

Figure 2. Total I(V) curve (in absolute value) for the symmetric (a) and asymmetric (b) cases in dark and under different illumination conditions.

The system under study is a single Si QD embedded in SiO_2 under illumination connected to two electrodes with a constant external bias voltage applied. We present two scenarios: the QD symmetrically connected $d = d'$ to the leads and in asymmetric configuration $d \neq d'$.
The obtained photocurrent as a function of the energy of the incident photons for different external bias voltages for the symmetric and asymmetric cases are shown in Fig. 1(a-b), respectively. For the V=0 case, in the symmetrically coupled system (a), the current is zero since the incoming hole currents for each side equal the outgoing electron currents. This analysis

Journal of Physics: Conference Series **609** (2015) 012002 doi:10.1088/1742-6596/609/1/012002

is inferred from the sum of all current terms; for the electron energy levels $I_{opt} - I_{Le} - I_{Re} = 0$ and for the hole levels $I_{Lh} + I_{Rh} - I_{opt} = 0$; where I_{opt} is the optical flux created among the electron/hole energy levels. I_{Le} and I_{Re} are the electron fluxes whereas I_{Lh} and I_{Rh} are the hole fluxes from the left and right leads, respectively. For the symmetrical coupling to both leads, $I_{Le} = I_{Re}$ and $I_{Lh} = I_{Rh}$. Therefore, $I_{Le} = I_{Lh}$ and the total current is zero since the hole currents compensate the electron ones. For the asymmetrical case (b), $I_{Le} \neq I_{Lh}$ and a net photocurrent is generated in the V=0 case.
When an external bias voltage is applied, the transmissions coefficient between the QD and the two leads change. Thus, the system becomes asymmetric and a net current appears. The current peaks are related to the maximum transition probabilities for an incident photon reflecting the absorption spectra. When the voltage increases, the current tends to be independent of the incident photon energy.
In Fig. 2(a-b), the obtained current voltage curves I(V) under external illumination as a function of the incident photon energies are shown for the symmetric and asymmetric cases. For the illuminated case, the main differences appear for small voltages when the optical terms dominates in Eq. 1 and the electrical response of the system differs from the dark case. Recovering the dark trend for higher voltages, when the current terms in Eq. 1 dominate again. Thus, the electrical response of the system is a competition between the two processes: the pure light current term and the external bias voltage term.

4. Conclusions
In conclusion, we present a theoretical framework based on non-coherent rate equations combined with the Transfer Hamiltonian approach to explain the transport properties of the next generation of optoelectronic devices based on the QDs properties. From a simple example based on a single Si QD embedded in SiO_2, we have demonstrated that the QD tunneling couplings to the leads play an important role in the final photoresponse of the system, being zero the net photocurrent when the QD is symmetrically coupled to he leads. The photocurrent response recover the shape of the absorption spectra. Concerning the I(V) curves in dark and illumination conditions, we have shown that the main differences appear due to the ratio between the optical and electrical terms.

References

[1] Illera S, Prades J D, Cirera A and Cornet A 2012 *EPL (Europhysics Letters)* **98** 17003 URL `http://stacks.iop.org/0295-5075/98/i=1/a=17003`
[2] Illera S, Prades J D, Cirera A and Cornet A 2012 *ArXiv e-prints* (*Preprint* 1207.5513)
[3] Illera S, Garcia-Castello N, Prades J D and Cirera A 2012 *Journal of Applied Physics* **112** 093701 (pages 7) URL `http://link.aip.org/link/?JAP/112/093701/1`
[4] Garcia-Castello N, Illera S, Guerra R, Prades J D, Ossicini S and Cirera A 2013 *Phys. Rev. B* **88**(7) 075322 URL `http://link.aps.org/doi/10.1103/PhysRevB.88.075322`
[5] Payne M C 1986 *Journal of Physics C: Solid State Physics* **19** 1145 URL `http://stacks.iop.org/0022-3719/19/i=8/a=013`
[6] Averin D V, Korotkov A N and Likharev K K 1991 *Phys. Rev. B* **44**(12) 6199–6211 URL `http://link.aps.org/doi/10.1103/PhysRevB.44.6199`
[7] Datta S *Quantum Transport: Atom to Transistor* (Cambridge University Press)
[8] Apalkov V M 2007 *Phys. Rev. B* **75**(3) 035339 URL `http://link.aps.org/doi/10.1103/PhysRevB.75.035339`
[9] Sakurai J J 1994 *Modern Quantum Mechanics Revised Edition* (Addison-Wesley Publishing Company)
[10] Melnikov D V and Fowler W B 2001 *Phys. Rev. B* **64**(24) 245320 URL `http://link.aps.org/doi/10.1103/PhysRevB.64.245320`
[11] Karabulut, afak H and Tomak M 2008 *Journal of Physics D: Applied Physics* **41** 155104 URL `http://stacks.iop.org/0022-3727/41/i=15/a=155104`
[12] Kayanuma Y 1988 *Phys. Rev. B* **38**(14) 9797–9805 URL `http://link.aps.org/doi/10.1103/PhysRevB.38.9797`
[13] Yu P Y and Cardona M 2005 *Fundamentals of Semiconductors* (Springer)
[14] Klingshirn C F 2007 *Semiconductor Optics* (Springer)

Journal of Physics: Conference Series **609** (2015) 012003 doi:10.1088/1742-6596/609/1/012003

Influence of dielectric environment on exciton and bi-exciton properties in colloidal, type II quantum dots

Jacek M. Miloszewski, Thomas Walsh, and Stanko Tomić

School of Computing, Science & Engineering, University of Salford, The Crescent, Salford, Lancashire M5 4WT, United Kingdom

Abstract. We present theoretical calculations of type II CdSe/CdTe quantum dots systems. We use an 8-band **k.p** Hamiltonian that includes spin-orbit interaction, strain, and first order piezoelectric effects. Exciton and bi-exciton states are found using the configuration interaction (CI) method that explicitly includes the effects of the Coulomb interaction, as well as exchange and correlation effects between many-electron configurations. We study convergence of the CI Hamiltonian with respect to the number of single particle states used in creation of the Hamiltonian. We show that there is a very strong correlation between the dielectric constant of the environment and exciton and bi-exciton energies.

1. Introduction

When a photon with energy greater than the energy of the bandgap is absorbed in a conventional solar cell, part of the energy (energy of the photon minus the energy of the bandgap) is always lost due to rapid thermalizaiton and Auger processes. However in a quantum dot solar cell part of this energy can be utilized due to the multiple exciton generation (MEG) process. In this process the high energy photon creates a high energy exciton that can decay into a bi-exciton (shown in Fig.1a). To observe MEG, the energy of the exciton (which usually consists of a high energy electron and a hole in the ground state) has to be at least twice as big as the energy of the effective optical gap $(E_{e_n} - E_{h_0} \geq 2\,(E_{e_0} - E_{h_0}))$, where e_0 and h_0 are the conduction band minimum and valence band maximum. This allows for a full utilization of high energy photons and dramatically increases solar cell efficiency. The MEG process competes with other radiative and non-radiative recombination processes (such as phonon and Auger cooling [1]).

To design the most optimal quantum dots it thus necessary to know what affects exciton and bi-exciton energies and how. In this work we focus on convergence issues of the CI Hamiltonian and the influence of the environmental dielectric constant on the exciton and bi-exction energies.

2. Theory

Our approach to the calculation of exciton and bi-exciton energies consists of two major steps: solution to a single particle Hamiltonian, and application of the single particle states to create and solve the configuration interaction (CI) Hamiltonian.

In the first step we apply an 8-band $\mathbf{k} \cdot \mathbf{p}$ Hamiltonian that includes continuum mechanical strain and piezoelectric effects [2, 3]. The Hamiltonian is parameterized using material

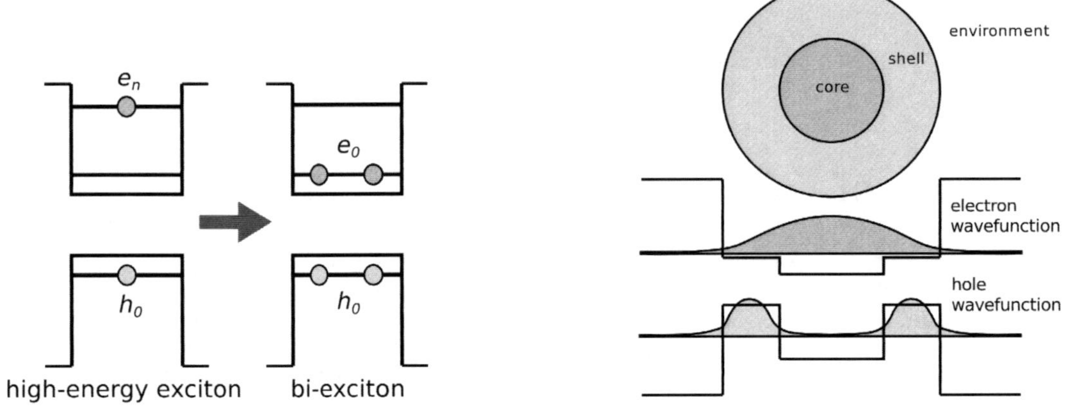

(a) (b)

Figure 1: (a) Representation of the multiple exciton generation process. An exciton that consists of a high energy electron e_n and valence band maximum hole h_0 decays into a bi-exciton state that consists of 2 conduction band minimum states e_0 and 2 valence band maximum states h_0. (b) Structure of a CdSe/CdTe core/shell type II quantum dot. In CdSe/CdTe core/shell QDs hole wavefunctions are confined in the shell region.

parameters taken either from experiment or from ab initio hybrid density functional calculations using B3LYP and PBE0 functionals.

Because of the large difference in dielectric permittivities, each charge inside the dot induces significant surface charges which strongly interact with the charges inside the dot. This is known as the surface polarization effect [4, 5]. The exact form of the Coulomb potential can be obtained by directly solving the Poisson equation, resulting in two terms: $V_{\mathrm{P}}(\mathbf{r}_1, \mathbf{r}_2)$, the two-particle interaction term, and $V_{\mathrm{SP}}(\mathbf{r})$, the self-polarization potential originating from the interaction of a charge with its own induced surface charge. The full potential is

$$V(\mathbf{r}_1, \mathbf{r}_2) = V_{\mathrm{P}}(\mathbf{r}_1, \mathbf{r}_2) + V_{\mathrm{SP}}(\mathbf{r}_1) + V_{\mathrm{SP}}(\mathbf{r}_2). \tag{1}$$

To speed up calculations, in this work we use the following approximation for the two-particle interaction

$$V_{\mathrm{P}}(\mathbf{r}_1, \mathbf{r}_2) \approx V_{\mathrm{P}}(|\mathbf{r}_1 - \mathbf{r}_2|) = \frac{e}{4\pi\epsilon_0\epsilon(|\mathbf{r}_1 - \mathbf{r}_2|)\,|\mathbf{r}_1 - \mathbf{r}_2|}, \tag{2}$$

where $\epsilon(r) = \epsilon_{\mathrm{dot}}$ for $r < R_{\mathrm{b}}$ and $\epsilon(r) = \epsilon_{\mathrm{env}}$ for $r > R_{\mathrm{b}}$. The self-polarization term $V_{\mathrm{SP}}(\mathbf{r})$ is calculated using the method developed by Bolcatto et al. [6]. The many-body Hamiltonian contains only particle-conserving terms, and is given by

$$\begin{aligned} H &= \sum_i E_i \hat{e}_i^\dagger \hat{e}_i - \sum_i E_i \hat{h}_i^\dagger \hat{h}_i + \frac{1}{2}\sum_{ijkl} V_{iljk} \hat{e}_i^\dagger \hat{e}_j^\dagger \hat{e}_k \hat{e}_l \\ &+ \frac{1}{2}\sum_{ijkl} V_{iljk} \hat{h}_i^\dagger \hat{h}_j^\dagger \hat{h}_k \hat{h}_l - \sum_{ijkl} (V_{iljk} - V_{ikjl})\, \hat{e}_i^\dagger \hat{h}_j^\dagger \hat{h}_k \hat{e}_l, \end{aligned} \tag{3}$$

where Coulomb integrals V_{iljk} are defined as

Journal of Physics: Conference Series **609** (2015) 012003 doi:10.1088/1742-6596/609/1/012003

$$V_{iljk} = \sum_{b=1}^{8} \sum_{b'=1}^{8} \int d^3\mathbf{r} \int d^3\mathbf{r}' \psi_b^{(i)} (\mathbf{r})^* \psi_b^{(l)} (\mathbf{r}) V (\mathbf{r}, \mathbf{r}') \psi_{b'}^{(j)} (\mathbf{r}')^* \psi_{b'}^{(k)} (\mathbf{r}') . \quad (4)$$

3. Results

First we focus on convergence issues of the CI Hamiltonian. Figs. 2a and 2b show convergence of bi-exciton energy (Exx) as a function of the number of single particle states for two different types of quantum dot: a CdSe core only QD and a CdSe/CdTe QD. Fig. 2a shows convergence of the CdSe type I QD with a fixed number of valence band states and a varying number of conduction band states and vice-versa. One can see that to reach a reasonable convergence it is enough to use 12 conduction band states and 20 valence band states. The analogous test for the case of the CdSe/CdTe type II quantum dot (Fig. 2b) indicates that convergence can be reached faster using only 10 conduction band states and 12 valence band states.

Next we focus on the influence of the environmental dielectric constant on the exciton and bi-exciton energies. In Fig. 3a we show the difference in energies (of excitons and bi-excitons) between two identical systems immersed in environments with $\epsilon_{\rm env} = 2.0$ and $\epsilon_{\rm env} = 1.9$, i.e. $\Delta Exx = |Exx_{1.9} - Exx_{2.0}|$ and $\Delta Ex = |Ex_{1.9} - Ex_{2.0}|$. One can immediately see that even a very small difference in the dielectric constant of the environment has a profound impact on the exciton and bi-exciton energies by approximately 15 and 30 meV respectively. In addition, this impact becomes even more pronounced for bi-excitons when the structure becomes type II, which happens for shell thicknesses larger than 4Å.

Finally we calculate exciton and bi-exciton energies for the same CdSe/CdTe core/shell QD for ten different values of environmental dielectric constant. In Fig. 3b one can see rapid increase of both exciton and bi-exciton energies with increasing value of the dielectric constant of the environment. The electron-hole binding energy is much larger for small values of $\epsilon_{\rm env}$ (when the ratio of $\epsilon_{\rm QD}$ to $\epsilon_{\rm env}$ is the biggest). For small values of $\epsilon_{\rm env}$ the polarization part of the Coulomb potential (1) dominates [7], hence the polarization effect increases the binding energy of electron-hole pairs.

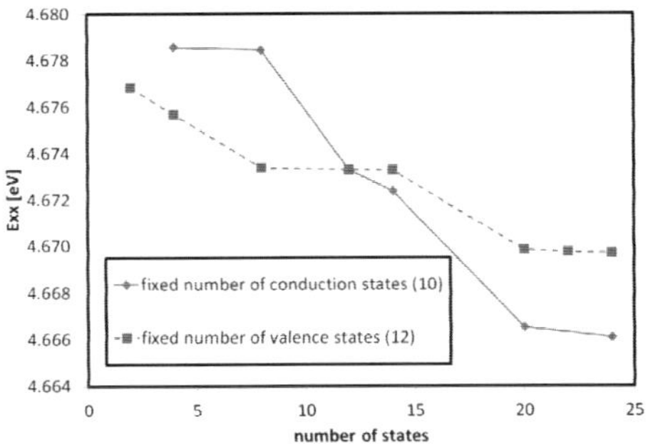

(a)

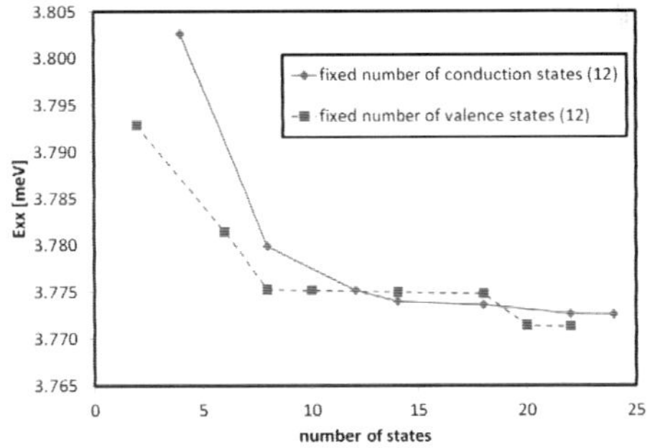

(b)

Figure 2: Convergence of bi-exciton energy with increasing number of states in CI Hamiltonian. (a) CdSe core only (type I) quantum dot with 15Å radius. (b) CdSe/CdTe core/shell structure (type II) with CdSe core (15Å radius) and 5Å thick CdTe shell.

4. Summary and conclusions

In conclusion, we calculated exciton and bi-exciton energies in type I CdSe and type II CeSe/CdTe colloidal quantum dots for various values of the environmental dielectric constant

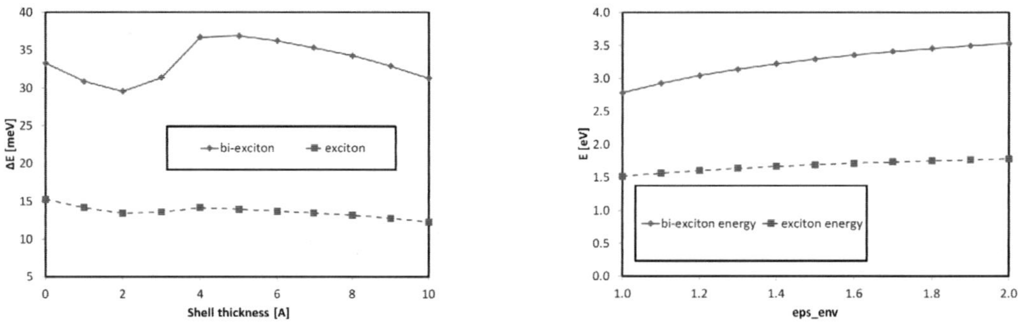

(a) (b)

Figure 3: (a) Difference in excition and bi-exciton energies for identical type II QDs with $\epsilon_{env} = 2.0$ and $\epsilon_{env} = 1.9$. (b) Exciton and bi-exciton energies of QDs (CdSe core with 20Å radius and 5Å thick shell) as a function of environmental dielectric constant..

and varying shell thickness. We showed that to get convergence of 1 meV it is necessary to take at least 20 valence band and 12 conduction band states (20/12) for type I QDs and 12/10 for type II QDs. We showed that the environmental dielectric constant has a significant impact on both exciton and bi-exciton energies, becoming even greater for type II QDs. Finally we proved that polarization effect increases the binding energy of excitons and bi-excitons and that this effect is even more pronounced for type II QDs.

5. Acknowledgements

This research is funded by: EPSRC grant "Enhanced multiple exciton generation in colloidal quantum dots" and the Royal Society grant "High Performance Computing in Modeling of Innovative Photo-Voltaic Devices". We acknowledge help from the N8 Research Partnership and Science and Technology Facilities Council for providing the computational resources used to conduct this research. We would also like to express our gratitude to Prof. Wendy Flavell, Prof. Paul O'Brien, and Prof. David Binks for useful discussions.

References

[1] Califano M, Zunger A and Franceschetti A 2004 *Applied Physics Letters* **84** 2409–2411 ISSN 0003-6951, 1077-3118
[2] Tomić S, Sunderland A G and Bush I J 2006 *Journal of Materials Chemistry* **16** 1963–1972 ISSN 1364-5501
[3] Vukmirovć N and Tomić S 2008 *Journal of Applied Physics* **103** 103718–103718–12 ISSN 00218979
[4] Piryatinski A, Ivanov S A, Tretiak S and Klimov V I 2007 *Nano Letters* **7** 108–115 ISSN 1530-6984
[5] Franceschetti A, Williamson A and Zunger A 2000 *The Journal of Physical Chemistry B* **104** 3398–3401 ISSN 1520-6106
[6] Bolcatto P G and Proetto C R 2001 *Journal of Physics: Condensed Matter* **13** 319 ISSN 0953-8984
[7] Miloszewski J M, Tomić S and Binks D 2014 *Journal of Physics: Conference Series* **526** 012005

Impact of oxide thickness on the performance of a GaAs NWFET

A Price and A Martinez
College of Engineering, Swansea University, UK
E-mail: A.C.PRICE.625036@swansea.ac.uk

Abstract. Nanowire field effect transistors (NWFETs) are being considered as a replacement for planar MOSFETs at the end of the ITRS. In this work, the change in the relative position of the valleys due to the oxide thickness is investigated, and the subsequent effect this has on the drain current in each valley. An NWFET of cross-section 2.2×2.2 nm^2 and channel length 6 nm has been considered. The strong confinement at this scale can cause the low mass Γ-valley to become raised in energy, such that it is higher than the heavy mass L and X-valleys. This results in very low current in the device. It was found that increasing the oxide from 0.2 nm to 0.8 nm caused the Γ-valley to lower in energy, resulting in a higher current. Both ballistic and dissipative transport are considered. This was achieved through the use of the non-equilibrium Green's function (NEGF) formalism. The mode space approach in the effective mass approximation was deployed, and the conduction band masses were extracted from tight binding simulations. Scattering was found to cause a 65%, 44% and 36% in the current for a 0.2 nm, 0.8 nm and 1.6 nm oxide thickness respectively.

1. Introduction

Nanowire field effect transistors are of interest due to their strong electrostatic control, and are being investigated as an option for the end of the International Technology Roadmap for Semiconductors (ITRS) [1]. Recent advances have led to the investigation of sub-10 nm gate lengths. In this work, we consider gate-all-around NWFETs of cross-section 2.2×2.2 nm^2 and of channel length 6 nm (see figure 1 (a)). The simulated NWFET is n-channel and has a [100] GaAs core. The source and drain are 15 nm long, and are doped at 10^{20} cm^{-3}. The channel is undoped.

III-V nanowires such as GaAs are relevant because of their high electron mobility in comparison to Si. In bulk, the order of valleys in increasing energy are Γ, L and X (figure 1 (b)). In previous work on 2.2×2.2 nm^2 cross-section gate-all-around NWFETs with a 0.8 nm oxide and 0.2 nm penetration of the wavefunction into the oxide, we found that the Γ-valley becomes raised in energy, such that it is higher than the L and X-valleys [2]. This is due to the strong confinement and the relatively low mass of the Γ-valley in comparison to the heavier mass L and X-valleys. In this work, we investigate the effect of increasing the thickness of the oxide and the penetration of the wavefunction into the oxide. We consider a 0.2 nm, 0.8 nm and 1.6 nm oxide thickness, with full penetration of the wavefunction into the oxide.

The Non-equilibrium Green's function (NEGF) approach is a powerful quantum transport formalism, which has been widely used to simulate nanoscale devices where quantum effects are in existence [3, 4]. In this work, we use the NEGF formalism within the effective mass

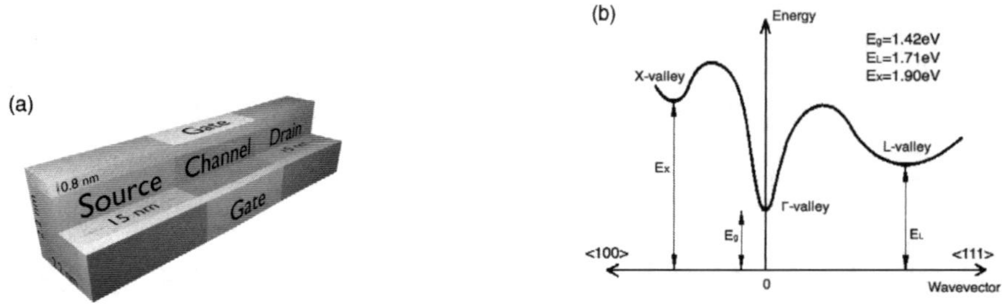

Figure 1. (a) A gate-all-around NWFET with a 2.2×2.2 nm^2 cross-section. (b) Bandstructure for bulk GaAs.

approximation (EMA). The effective masses have been extracted from tight binding simulations [5]. Phonon scattering has been introduced through the use of the self-consistent Born approximation.

2. Theory

The NEGF approach describes the modelled system by using the one-particle electronic Green's functions, G, and the self-energies, Σ. The retarded Green's function is calculated using

$$G^r(E) = [E1 - H - \Sigma^r(E)]^{-1} \tag{1}$$

where H is the Hamiltonian of the system. The lesser-than, $G^<$, and the greater-than, $G^>$, Green's functions are given by

$$G^{\lessgtr}(E) = G^r(E)\Sigma^{\lessgtr}(E)G^a(E) \tag{2}$$

The advanced Green function, G^a, is the complex conjugate of the retarded Green's function, G^r. The retarded self-energy, Σ^r, is given by

$$\Sigma^r(E) = \frac{1}{2}(\Sigma^<(E) - \Sigma^>(E)) \tag{3}$$

where $\Sigma^<$ and $\Sigma^>$ are the lesser-than and greater-than self-energy respectively.

3. Results

The effect of increasing the oxide thickness is investigated by considering firstly the penetration of the wavefunction for the different oxide thicknesses; the subsequent effect this has on the relative position of the valleys is then investigated. The I_D-V_G characteristics for every simulated NWFET, at both low and high drain bias are included, demonstrating the effect of the oxide thickness on the drain current. Finally, the I_D-V_G characteristics for each valley at low drain bias are included, showing how the current in each valley changes with differing oxide thickness.

The effect of the increasing oxide penetration on the transverse potential energy profile can be observed in figure 2. For the 0.2 nm oxide thickness in figure 2 (a), the potential energy linearly increases as it penetrates the oxide. For the other two cases (b) 0.8 nm thickness and (c) 1.6 nm thickness, the potential energy levels off, after it's penetrated 0.2 nm into the oxide. The potential energy profile shifts upwards as the oxide penetration increases. The penetration of the wavefunction increases as the oxide penetration increases.

Figure 3 shows the energy-resolved current spectra for the (a) 0.2 nm, (b) 0.8 nm and (c) 1.6 nm oxide thickness. For the 0.2 nm oxide, the Γ-valley is raised in energy, such that it's

Journal of Physics: Conference Series **609** (2015) 012004 doi:10.1088/1742-6596/609/1/012004

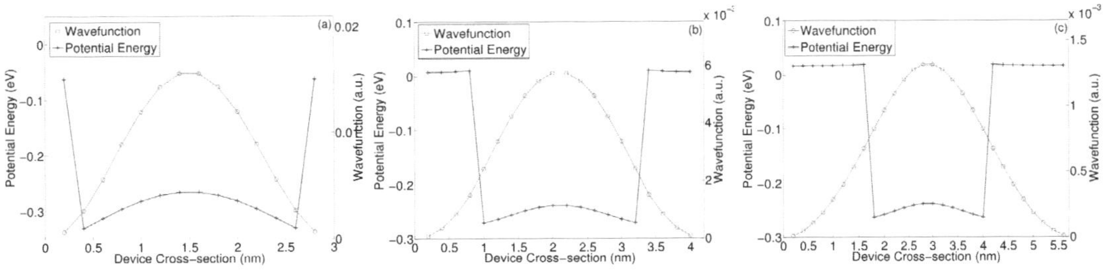

Figure 2. Transverse potential energy profile and Γ-valley wavefunction for (a) 0.2 nm, (b) 0.8 nm and (c) 1.6 nm oxide thickness.

above the L and X-valley. This is due to the low mass of the Γ-valley in comparison to the heavier L and X-valleys. As the oxide thickness increases, the Γ- valley lowers in energy, such that it is below the L and X-valleys. For the 0.2 nm, 0.8 nm and 1.6 nm oxide thickness, the Γ-valley is 0.65 eV, 0.33 eV and 0.28 eV higher than the potential (denoted by the magenta line) respectively. The subsequent effect this has in the current can be seen in figure 4 (a) and (b), which show the I_D-V_G characteristics at low and high drain bias respectively. As the oxide thickness increases, the drain current increases.

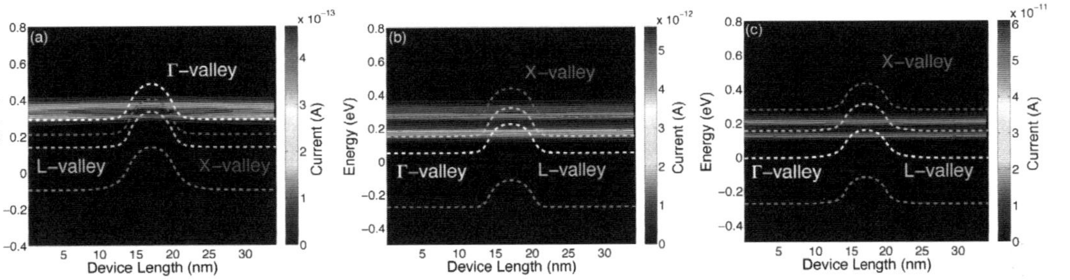

Figure 3. Energy-resolved current spectra for (a) 0.2 nm, (b) 0.8 nm and (c) 1.6 nm oxide thickness (the magenta dotted line is the potential energy profile).

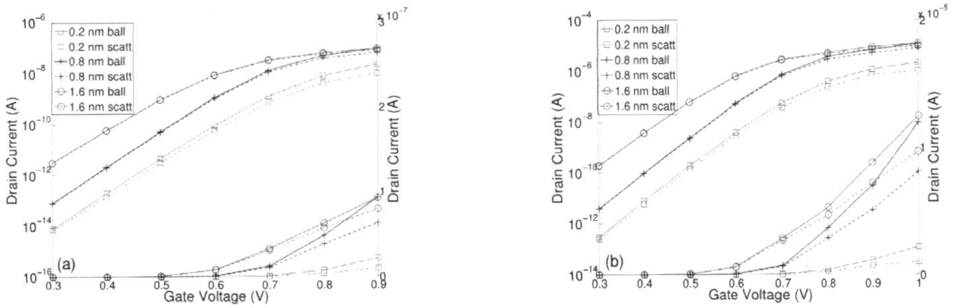

Figure 4. I_D-V_G characteristics at (a) low drain bias, V_D = 1 mV, and (b) high drain bias, V_D = 0.6 V.

Figure 5 breaks down the I_D-V_G characteristics for each valley at low drain bias for the (a) 0.2 nm, (b) 0.8 nm and (c) 1.6 nm oxide thickness. As the oxide thickness increases, the relative amount of current for the L and X-valleys remains similar, but the relative amount of current in the Γ-valley increases dramatically. For the 0.2 nm oxide thickness, the heavy mass L-valley has the greatest current. However, for the 0.8 nm and 1.6 nm oxide thickness the low mass Γ-valley has the greatest current.

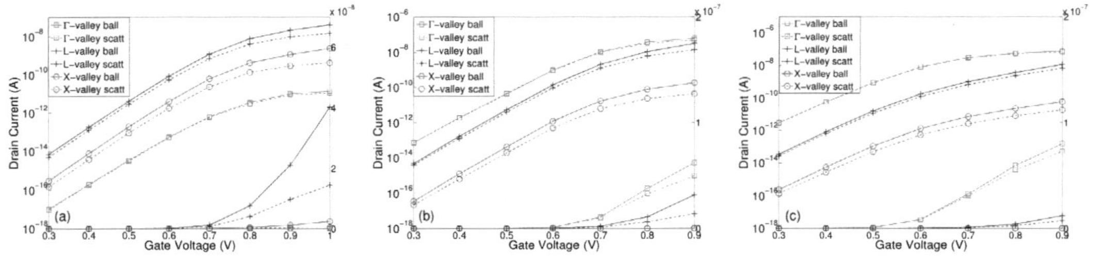

Figure 5. I_D-V_G characteristics per valley at low drain bias for (a) 0.2 nm, (b) 0.8 nm and (c) 1.6 nm oxide thickness.

The increase in oxide thickness also effects the reduction in the drain current due to scattering. As the oxide thickness increases, the effect of scattering is reduced. Scattering causes a 65%, 44% and 36% reduction in the on-current for the 0.2 nm, 0.8 nm and 1.6 nm oxide thickness respectively.

4. Conclusions

In previous work on a gate-all-around 2.2×2.2 nm^2 cross-section NWFET, it was found that the strong confinement caused the low mass Γ-valley to become raised in energy such that it was higher than the heavier L and X-valleys. In this paper, the effect of increasing the thickness of the oxide was investigated. Three different oxide thicknesses were considered: 0.2 nm, 0.8 nm and 1.6 nm. It was found that for the 0.2 nm oxide thickness the Γ-valley was above the L and X-valleys, but as the penetration of the wavefunction increased to 0.8 nm the Γ-valley was then below the L and X-valleys. For the 1.6 nm oxide thickness, the Γ-valley was lowered slightly further. For the 0.2 nm, 0.8 nm and 1.6 nm oxides the Γ-valley was 0.65 eV, 0.33 eV and 0.28 eV higher than the potential respectively.

This had a subsequent effect on the current, with the drain current increasing as the oxide increases. This result can be understood by studying the I_D-V_G characteristics for each valley. As the oxide thickness increases, the drain current in the L and X-valley changes very little. But as the penetration into the oxide increases from 0.2 nm to 0.8 nm the drain current in the Γ-valley increases by the order 10^4. The increase in the oxide thickness also reduces the impact of scattering on the drain current. For the 0.2 nm, 0.8 nm and 1.6 nm oxide penetration, there was a 65%, 44% and 36% reduction in the on-current respectively.

References

[1] http://www.itrs.net/

[2] A. Price and A. Martinez, "Investigation on phonon scattering in a GaAs nanowire field effect transistor using the Non-Equilibrium Greens Function formalism" (2014, unpublished).

[3] R. Lake and S. Datta, *Phys. Rev. B* **45**, 6670 (1991).

[4] M. Aldegunde, A. Martinez and A. Asenov, *J. Appl. Phys.* **110**, 094518 (2011).

[5] Y. M. Niquet, A. Lherbier, N. H. Quang, M. V. Fernandez-Serra, X. Blase and C. Delerue, *Phys. Rev. B* **73**, 165319 (2006).

Modelling of Thermal Boundary Resistance in a GaN Diode by means of Electro-Thermal Monte Carlo Simulations

S García, I Íñiguez-de-la-Torre, Ó García-Pérez, J Mateos, T González and S Pérez
Departamento de Física Aplicada, Universidad de Salamanca
Plaza de la Merced, s/n, 37007 Salamanca, Spain.

e-mail: sergio_gs@usal.es

Abstract. In this paper we evaluate heat diffusion in an AlGaN/GaN diode through a Monte Carlo simulator by expanding its capabilities with the implementation of two thermal methods. We present the impact on the device temperature of considering different substrates and die dimensions. We also evaluate the influence of the thermal boundary resistance (TBR) that appears in the growth process of dissimilar materials. We analyse the effect of the TBR when the diode is grown on two substrates, Si and SiC. As a conclusion, we can state that the TBR is a limiting factor to the thermal flow that becomes more relevant for substrates with high thermal conductivities.

1. Introduction

Self-heating properties are an important problem for high-power gallium nitride (GaN) electronic and opto-electronic devices because thermal effects can cause strong damages and premature failure. These effects are boosted by the TBR. The TBR is a measure of the resistance that an interface presents to the thermal flow due to different phonon dynamics and poor crystalline quality near the boundary. The TBR can be measured by employing 3-D micro-Raman thermography [1]. In order to include TBR effects in semiconductor device simulators, different techniques are used, from those based on 3-D finite-difference models [1] to others that employ a continuity condition for a finite interfacial conductance between the layers of interest [2]. But until nowadays the modelling of the TBR effects is still an open problem. In this work, thermal effects will be studied by means of two thermal approximations, both of them implemented in our electronic home-made Monte Carlo (MC) simulator. One is based on a simple Thermal Resistance Method (TRM) [3],[4],[5]. The other is based on the solution of the steady state heat-diffusion-equation (HDE) [4],[6].

This paper is organized as follows. In Sec. 2 the geometry and material components are described followed by a brief explanation of the used electro-thermal models. In Sec. 3 we show the main results for different substrates and die lengths of the diode, making emphasis on the effect of the inclusion of the TBR. Finally, Sec. 4 summarizes the main conclusions.

2. Device structure and electro-thermal simulations

In Figure 1 the geometry of the $Al_{0.27}Ga_{0.73}N/GaN$ diode under analysis is presented. The electronic transport will take place in the called electronic domain (delimited by red dashed lines in Figure 1). Piezoelectric scattering and those with phonons and dislocations are included in the model. The influence of spontaneous and piezoelectric surface polarization charges P=12.12×10^{12} cm^{-2} have been incorporated. In addition, a surface charge density σ = - 4.12×10^{12} cm^{-2} is placed at the top of the

AlGaN layer. The length of the diode is 2 µm. If not indicated, L_1=200 µm and L_2=300 µm. In this work, temperature-independent thermal conductivities will be employed: k_{AlGaN}=30 W/(K·m) [7], k_{GaN}=130 W/(K·m) [8], k_{Au}=300 W/(K·m) [6], $k_{diamond}$=1000 W/(K·m) [9], k_{Si}=156 W/(K·m) [6], k_{SiC}=300 W/(K·m) [10], and $k_{sapphire}$=42 W/(K·m) [11]. A more detailed analysis for temperature-dependent thermal conductivities can be found in Ref. [4].

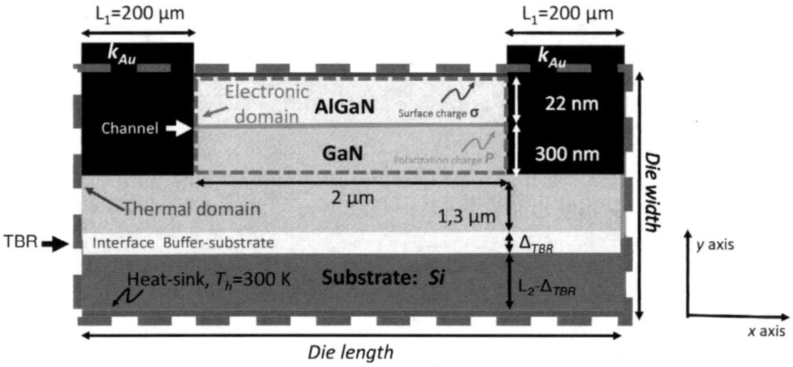

Figure 1. Structure of the $Al_{0.27}Ga_{0.73}N$/GaN diode under study.

Based on our home-made ensemble MC simulator coupled with a 2D Poisson solver [5], we have expanded its capabilities by including the two thermal approaches [4] sketched in Figure 2.

- (*i*) *Thermal Resistance Method* (TRM) (blue colour in Figure 2). This model is based on the use of an ad-hoc thermal resistance, R_{th}, and is only carried out in the electronic domain (delimited by red dashed lines in Figure 1). The lattice temperature is updated every N_{th}=5000 iterations according to the formula:

$$T_{latt}=300\text{ K} + P_{diss} \times R_{th} \ , \tag{1}$$

  with P_{diss},$=I_{DC} \times V_{DC}$ the dissipated power in the N_{th} iterations. The advantage of this model to evaluate the thermal behaviour of devices is the simplicity and low computational effort with respect to other algorithms.

- (*ii*) *Solution of the* HDE (we refer from now to this model HDEM, orange colour in Figure 2). In this case the lattice temperature is updated according to the solution of the HDE equation:

$$\nabla[k(r)\nabla T(r)] = -G(r) \ , \tag{2}$$

  where $k(r)$ is the temperature-independent and inhomogeneous thermal conductivity, T the temperature, and $G(r)$ the power density distribution generated by phonons. Note that the HDE is solved in the whole thermal domain (delimited by green dashed lines in Figure 1). In this case, the HDE is solved every N_{th}=10000.

Boundary conditions are extremely relevant in this kind of simulations. To correctly model the interfaces between layers, we use the continuity condition:

$$k_1\, \partial T/\partial r_n|_\gamma = k_2\, \partial T/\partial r_n|_\gamma, \tag{3}$$

where γ is the interface between two layers with thermal conductivities k_1 and k_2, and r_n the normal position vector. At the two lateral edges, and at the top of the device, adiabatic boundary conditions are imposed, $k_1\, \partial T/\partial r_n|_\gamma = 0$. At the bottom, a Dirichlet boundary condition is used; a heat sink at 300 K. Our approach to take into account the effect of *TBR* is very simple: we include a small layer of thickness Δ_{TBR} of a material with low thermal conductivity k_{TBR} between the GaN-buffer and the substrate (see Figure 1, yellow layer). In this framework the value of the simulated TBR will be:

$$\text{TBR} = \Delta_{\text{TBR}}/k_{TBR}\ . \tag{4}$$

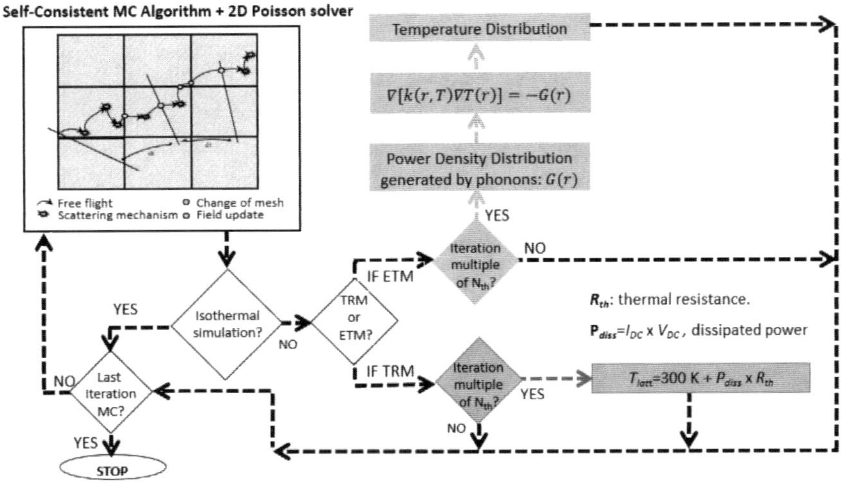

Figure 2. Flow chart of the electro-thermal simulations.

3. Main results

First of all, the effect of growing the device on different substrates is examined through the HDEM, but without the TBR layer. In Figure 3 we plot the *I-V* curves, Figure 3 (a), and the average temperature (T_{av}) in the electronic domain *vs.* the dissipated power, Figure 3 (b). The fitting of T_{av} *vs.* P_{diss} allows us to extract a thermal resistance R_{th}, which is in the range 3.5-38x10^{-3} K·m/W. Moreover, within the HDEM it is possible to study the dependence of that extracted R_{th} on the die dimensions. For example, the values of R_{th} *vs.* L_1 (for L_2 250 µm, 300 µm, and 350 µm) are shown for Si-substrate in the inset of Figure 3 (b). As L_2 is reduced, R_{th} becomes nearly constant with L_1 for lower values of L_1. For a given L_1, as L_2 is increased, the thermal resistance is higher, as expected, since the heat flux generated by the phonons is dissipated less efficiently due to the larger distance from the heat sink to the electronic domain. R_{th} is in the range 12-23x10^{-3} K·m/W.

Secondly, the effect of the TBR is analysed by choosing as reference two substrates, Si and SiC. A layer of thickness Δ_{TBR}=15 nm is introduced at the interface buffer-substrate. Simulations have been performed by considering two thermal conductivities (k_{TBR}) of 1 and 0.1 W/(K·m) for this layer. According to eq. (4) the TBR is equal to 1.5x10^{-8} and 15x10^{-8} m^{2}K/W, being consistent with the reported experimental [1] and numerical studies [12]. In Figure 4 (a) and (b), T_{av} *vs.* P_{diss} is analysed for both substrates. The insets show the profile of the lattice temperature in the middle of the structures of the interface buffer-substrate. As expected, SiC exhibits better thermal behaviour than Si, and lower temperatures are obtained inside the device. But if the TBR is very high (increasing the difficulty for the thermal flow to leave the device), it can be the critical parameter for the design of an appropriate heat sink. Note that the discontinuity ΔT at the interface increases when a higher conductivity substrate is chosen (as example for k_{TBR} =0.1 W/(K·m) ΔT of 125 K and 142 K are obtained for Si and SiC, respectively). Remarkably it may happen that a device grown on a substrate with poor k_s but with a good thermal interface resistance (low TBR) exhibits better thermal behaviour than other grown on a substrate with an excellent k_s but with a high TBR. This behaviour is also reflected in the values of R_{th} obtained, that are in the range 14-21x10^{-3} K·m/W, and 9-17x10^{-3} K·m/W, for Si and SiC, respectively.

Finally, we have proven that both thermal models are equivalent, i.e., a TRM simulation [symbols of Figure 3 (a)] with the thermal resistance extracted from the HDEM simulation (R_{th}=3.58x10^{-3} K·m/W for diamond, 8.9x10^{-3} K·m/W for SiC, 14.1x10^{-3} K·m/W for Si, and 38.2x10^{-3} K·m/W K·m/W for sapphire) provides the same *I-V* curve than HDEM. Obviously, once the TBR is included in the HDEM simulation, a TRM simulation (with the appropriate extracted R_{th}) would also provide the same result than the HDEM.

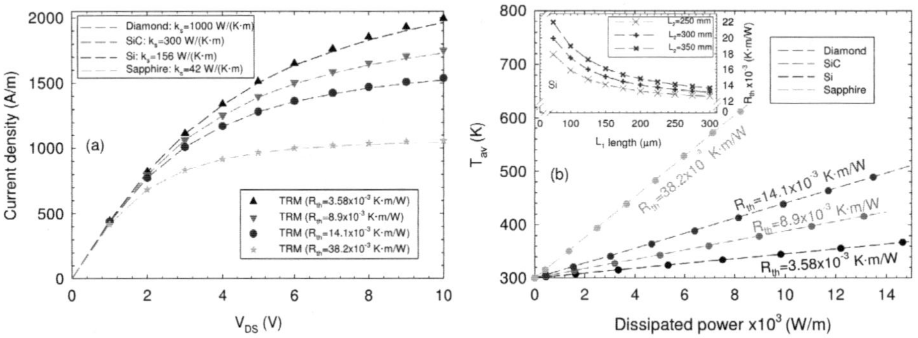

Figure 3. (a) *I-V* curves for different substrates obtained with the HDEM simulation. Symbols: TRM with R_{th}=3.58x10^{-3}, 8.9x10^{-3}, 14.1x10^{-3} and 38.2x10^{-3} K·m/W. (b) Average temperature, T_{av}, *vs.* the dissipated power, and linear fitting to extract the corresponding thermal resistance R_{th}. The inset shows R_{th} *vs.* die dimension L_1 for L_2=250 μm, 300 μm and 350 μm and Si-substrate. If not indicated, L_1=200 μm and L_2=300 μm.

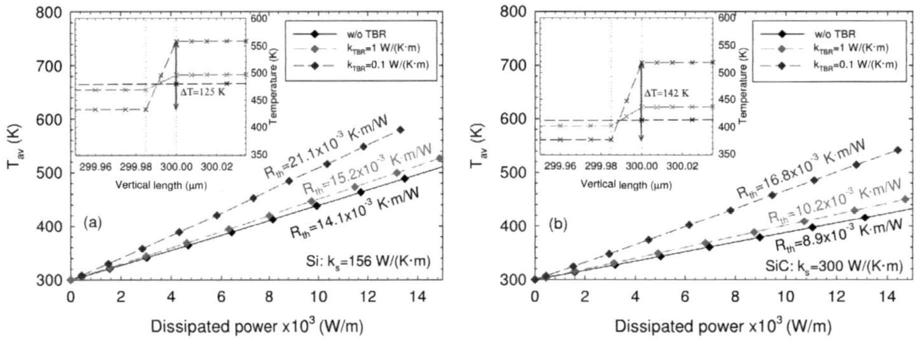

Figure 4. T_{av} *vs.* P_{diss} for two substrates (a) Si, and (b) SiC. A TBR layer of Δ_{TBR}=15 nm with k_{TBR} =1 and 0.1 W/(K·m) is studied. Note that the case without TBR layer from Fig. 3 are included for comparison. The inset shows the profile of the lattice temperature in the middle of the diode close to the TBR region, for a bias of 10 V.

4. Conclusions

In this paper we have presented the effect of the inclusion of the TBR in our MC simulator. Results show that the TBR may have a strong influence, limiting the thermal flux in the interface where it appears. Although the effect of the TBR is higher for the SiC substrate, the temperature reached in the electronic domain is lower compared to the temperature obtained with the Si substrate. The TBR has a strong influence on R_{th}. For a TBR=15x10^{-8} m^{2}K/m, R_{th} is increased respect the case w/o TBR a factor 1.5 for Si and 1.89 for SiC.

Acknowledgments: This work has been partially supported by the Dirección General de Investigación Científica y Técnica (MINECO) through the Project TEC2013-41640-R and by the Consejería de Educación de la Junta de Castilla y León through the Project SA052U13.

[1] Sarua A, Hilton K P, Wallis D J, Uren H M, et al. *IEEE Trans Electron Devices* **54** 3152-3158
[2] Muzychka Y S, Bagnall K R, and Wang E N, *IEEE Trans Compon Packag Manuf Technol* **3** 1826-1841.
[3] Asgari A, Kalafi M, and Faraone L. *Physica E-Low-Dimensional Systems & Nanostructures* **28** 491-499.
[4] García S, Íñiguez-de-la-Torre I, García-Pérez Ó, et al. Semicond. Sci. Technol. 30 035001.
[5] Mateos J, Pérez S, Pardo D and González T. *Conf. on Electron. Devices IEEE Catalog CFP* 459-462.
[6] Bonani F, and Ghione G. *Solid-State Electronics* **38** 1409-1412.
[7] Liu W L, Balandin A A. *Appl. Phys. Lett* **85** 5230-5232.
[8] Vitanov S, Palankovski V, Maroldt S, and Quay R. *Solid-State Electronics* **54** 1105-1112.
[9] Wang A, Tadjer M, and Calle F. *Semicond. Sci. Technol.* **28** 055010.
[10] Tang X, Rousseau Michel, Defrance N, et al. *Phys. Status Solidi A* **207** (8) 1820-1826.
[11] Sadi R, Kelsall R W, and Pilgrim N J. *IEEE Trans. Electron Devices* **53** 2892-2900.
[12] Filippov K, and Balandin A A. *MRS Internet J. Nitride Semiconductor Research* [*online*] **8** article 4.

Complete collisions approximation to the Kadanoff–Baym equation: a first–principles implementation.

Davide Sangalli
CNR-ISM, U.o.s. di Montelibretti, via Salaria Km 29.3, I-00016 Montelibretti, Italy, European Union
E-mail: davide.sangalli@ism.cnr.it

Andrea Marini
CNR-ISM, U.o.s. di Montelibretti, via Salaria Km 29.3, I-00016 Montelibretti, Italy, European Union

Abstract. We show carriers dynamics on bulk Silicon in the sub pico–second time scale. The results are obtained from a first–principles implementation of the the Kadanoff–Baym equations within the generalized Baym–Kadanoff ansatz and the complete collision approximation. The resulting scattering term is similar to the scattering described within the semi–classical Boltzmann equation [1].

1. Introduction

The description of carriers dynamics in semi–conductors is crucial for a correct modelization of electronic devices. Silicon (Si) in particular is the fundamental building block of semi–conductors physics and microelectronics industry [2].

The reference equation used in the literature is the semi–classical Boltzmann equation (sc-BE) which describes the scattering of electrons and holes with lattice vibrations, electron–phonons (e–p) scattering, with other carriers, electron–electron (e–e) scattering and possibly with defects. Its implementation is usually based on models, where the characteristic of the system are taken into account with ad–hoc parameters. In the present work we want instead to describe such scattering events within a first–principles approach.

Also the sc–BE assumes a semi–classical regime, as opposite to the quantum–kinetic regime which should, in principle, describe the electrons propagation. However the miniaturization of Si–based devices to the nano–scale regime and the never ending search for faster devices call for a deep understanding of the fundamental quantum–mechanical process that governs the carriers dynamics [3, 4]. Moreover the recent development of ultra–short laser pulses has triggered the necessity to understand the quantum kinetics of electrons and holes on the sub pico–second (ps) time–scale.

The non–equilibrium Green's functions (NEGF) approach offers a quantum–kinetic equation, the Kadanoff–Baym equation (KBE), which describes the quantum dynamics through the many–body Green's function $G^{<}(t, t')$. The solution of the full equation however is not feasible within a first–principles approach, and also for simple models, an approximate version is usually adopted. The most common approximation is the generalized Baym–Kadanoff ansatz (GBKA), which reduces the KBE to a closed equation for the time–diagonal Green's function, i.e. the time–dependent density matrix of the system. If also the complete–collision approximation (CCA) is employed the scattering processes are described at the same level of the sc–BE [1], however with a smeared energy conservation due to possible violations on the ultra–short time regime. We will refer to this equation as the CCA–KBE.

The CCA–KBE has been recently implemented in the yambo [5] code. Projected onto the Kohn–Sham (KS) wave–functions, it describes the evolution of a parameter free $G^<_{ij}(t,t)$, which can be used to reconstruct the electronic occupations: $f_i = -iG^<_{ii}(t,t)$. Here we used the generalized index $i = \{n\mathbf{k}\sigma\}$. Although the scattering terms still have a semi–classical form, the CCA–KBE offers a suitable starting point for future development towards a real quantum kinetics equation. It also has other advantages. For example it can correctly model the interaction with ultra–short laser pulses. Indeed the first–principles CCA–KBE has been shown to reproduce the propagation of the polarization [6], $P = \sum_{i\neq j} G^<_{ij}(t,t)$ induced by a laser pulse.

In the present work we explore the performance of its implementation on bulk Silicon and in particular we will focus our attention of the scattering term $S_i(t)$. The details and the approximations needed for a first–principles implementation of this term have been discussed in Ref. [7].

2. The theoretical method

The theoretical approach used is described in Refs. [6, 7]. Here we report the main equations. The CCA–KBE equation reads

$$i\hbar\partial_t G^<_{ij}(t) - \left[H_{MB} + H_{RT}(t), G^<(t)\right]_{ij} = S_{ij}(t) \tag{1}$$

with

$$H_{DFT}[\rho^{eq}] = H_0 + V_H[\rho^{eq}] + V_{xc}[\rho^{eq}] \tag{2a}$$
$$H_{MB}[G^{<,eq}] = H_{DFT}[\rho^{eq}] + (\Sigma[G^{<,eq}] - V_{xc}[\rho^{eq}]) \tag{2b}$$
$$H_{RT}[G^<(t)](t) = V_H[\delta\rho(t)] + \left(\Sigma[\delta G^<(t)], G^{<,eq}]\right) + U^{ext}(t) \tag{2c}$$
$$S(t) = \{\Sigma^>_C, G^<\} + \{\Sigma^<_C, G^>\} \tag{2d}$$

H_{DFT} is the standard KS Hamiltonian. H_{MB} describes the quasi–particle (QP) corrections. Finally the H_{RT} Hamiltonian instead includes the external field and the variation in the Hartree potential and in the self–energy [6]. In the present work we replace its role with an ad–hoc initial guess for the electrons and holes populations.

The term $S(t)$ describes the carriers scattering [7]. Since the electronic occupations are the diagonal terms of the $G^<ij$, we focus our attention on the diagonal of eq. 1. The latter has only contributions from the $S(t)$ term, since we replace the role of H_{RT} with an initial guess and H_{MB} gives no contribution on the diagonal. Indeed the CCA–KBE, together with a QP approximation for the retarded and advanced Green's function [7], has the form

$$\partial_t f_i(t)|_{relax} = -\gamma^{(e)}_i(t) f^{(e)}_i(t) + \gamma^{(h)}_i(t) f^{(h)}_i(t), \tag{3}$$

with $f^{(e)}_i = f_i$ and $f^{(h)}_i = 1 - f_i$. The electron and hole lifetimes, $\gamma^{(e)}$ and $\gamma^{(h)}$, include both (e–p) and (e–e) contribution: $\gamma^{(e/h)} = \gamma^{(e/h)}\big|_{e-e} + \gamma^{(e/h)}\big|_{e-p}$. $\gamma^{(e/h)} = \gamma^{(e/h)}\big|_{e-e} + \gamma^{(e/h)}\big|_{e-p}$. In the e–e case we have that:

$$\gamma^{(e/h)}_i\Big|_{e-e} \propto \sum_{jkl} |W_{ij\to kl}|^2 P(\Delta_{ij} - \Delta_{kl}) f^{(h/e)}_j(t) f^{(h/e)}_k(t) f^{(e/h)}_l(t), \tag{4}$$

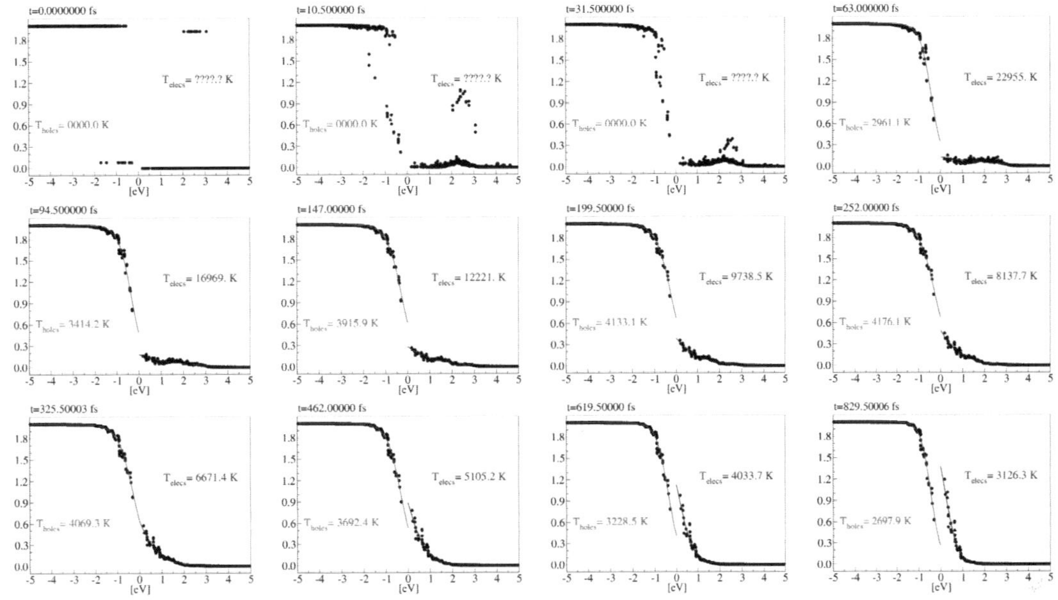

Figure 1. Carriers occupations distribution as a function of time

while in the e–p case

$$\gamma_i^{(e/h)}\Big|_{e-p} \propto \sum_{j\lambda}\sum_{I=\pm} N_I^\lambda \left|g_{i\to j}^\lambda\right|^2 P\left(\omega_\lambda \pm I\Delta_{ij}\right) f_j^{(h/e)}(t) . \quad (5)$$

In Eqs. 4-5 $\Delta_{ij} = \epsilon_i - \epsilon_j$ and λ represents a generic phonon mode with momentum $\mathbf{q}$ and branch η. $N_+^\lambda = (1 + N_\lambda(T)/2)$ and $N_-^\lambda = N_\lambda(T)/2$, with $N_\lambda(T)$ the Bose distribution function at energy ω_λ and temperature T. In the present work we assume that the phonons distribution does not change during the dynamics. W is the statically screened component of the Couloumb interaction whose frequency dependence is embodied in the P functions. These represent a smeared energy conservation condition. Finally, the g are the matrix elements of the screened ionic potential derivative, calculated within DFPT [8, 9].

3. Results on bulk Silicon

In the present work we used the local–density approximation (LDA) to describe bulk Silicon with norm–conserving pseudo–potentials with an energy cet–off of 20 Ha. For the Brilloine zone we use a uniform $4x4x4$ sampling and we consider the first 7 bands, i.e. 4 occupied bands and 3 empty bands. The QP corrections are then described with a scissor operator of 0.8 eV, which is known to be a very good approximation for buk Si. Then we compute the phonons and the e–p matrix elements with a DFPT run using the same parameters. The CCA–KBE is then propagated in the KS wave–functions space, but on an auxiliary $15x15x15$ k–grid. This bigger grid is required to ensure a proper description of the e–p scattering. The KS eigen–values are computed on this finer grid, while all other quantities are interpolated from the $4x4x4$ grid with a constant nearest–neighbor interpolation. The phonons are assumed to remain in equilibrium according to a Bose distribution of 300 K. The time step for the propagation is $dt = 10\ as$, while we use a second–order Runge–Kutta integration scheme. The conservation of the number of carriers during the dynamics is guaranteed by the detailed balance, as discussed in Ref. [7].

Journal of Physics: Conference Series **609** (2015) 012006 doi:10.1088/1742-6596/609/1/012006

We consider a starting point which mimics a high number of electrons excited from a laser pulse centered at the optical band–gap of Si. Thus we create near the Γ point, a packet of holes in valence and a packet of electrons in conduction. We use here a very high density of carriers, which would physically lead to a drastic change to the characteristics of the material. The choice is done to test the implementation in a high–carriers regime. Indeed here we are interested in exploring two selected physical effects, i.e. the e–e scattering and the e–p scattering, in this regime. In a lower density regime the effect of the e–e scattering would become smaller and it would be very difficult to test the reliability of the implementation and eventual errors.

In doing this we will neglect some physical effects, such as the softening of the phonon frequencies, the change in the band structure due to the extra carriers, the deviation of the phonons distribution from a Bose function and the sub–sequent heating of the lattice. All these effects becomes negligible in a lower density regime whereas instead the weaker e–e scattering would still have a role.

In Fig. 1 we see different time–snapshots of the carriers distribution as a function of the energy. We see that the electrons and the holes, which at the initial time are in a highly non–equilibrium situation, relax towards two Fermi distributions with changing temperatures. In the holes channel, carriers are closer to their minimum (i.e. the valence band maximum, VBM) and they reach a lower temperature Fermi distribution. In the electrons channel, instead, we can appreciate the different role played by the e–e scattering and the e–p scattering. Phonons have, on average, an energy of few meV and can scatter electrons only to levels energetically nearby, the result is a packet of electrons which keeps approximately the initial spread in energy and slowly moves towards the conduction band minimun (CBM). This is poorly visible because meanwhile the e–e channel is able to directly scatter electrons to the CBM.

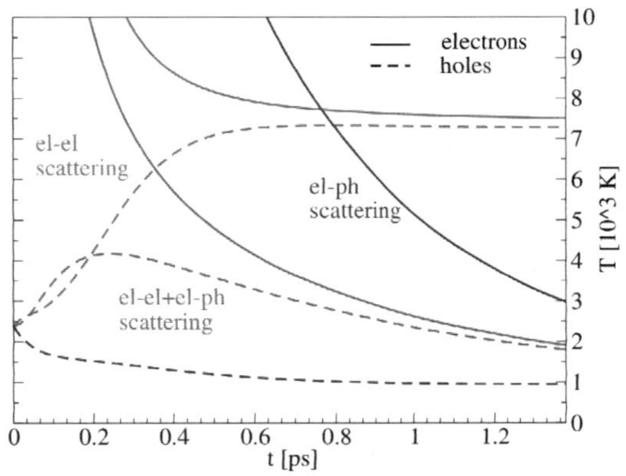

Figure 2. Overview of the electrons and holes temperatures for different scattering mechanisms

The resulting temperatures, obtained as the fit of the two Fermi distributions, are reported in Fig. 2. Here, together with the electron and hole temperature for the simulation with both e–e and e–p scattering on we report also the fit in case only one of the two relaxation channels is open. The e–e scattering (red lines) makes the electrons and holes interact and thus increase the holes temperature lowering the electrons temperature. However once the two sub–systems reach a thermal equilibrium nothing more can happen. Instead the e–p channel (blue lines) makes the carriers exchange energy with the lattice. Thus both electrons and holes decrease their temperature. However since phonons do not have enough energy to cross the electronic gap, the two processes are independent. Finally, when both scattering channels are open the resulting effect is a combination of the two. In this high carriers density regime, the e–e scattering dominates the dynamics in the initial part of the dynamics while, once electrons and holes reach a similar temperature, the e–p channel becomes the dominant channel.

The temperatures reached by electrons and holes depends mainly on the initial guess (or in case a laser pulse were used on its characteristics) and on the Silicon band structure. They are instead almost independent on the total number of carriers created. Indeed lowering the carriers density the main change is that the e–e scattering become less effective while all occupations numbers are rescaled. In the limit of very low carriers concentration the evolution of the carriers

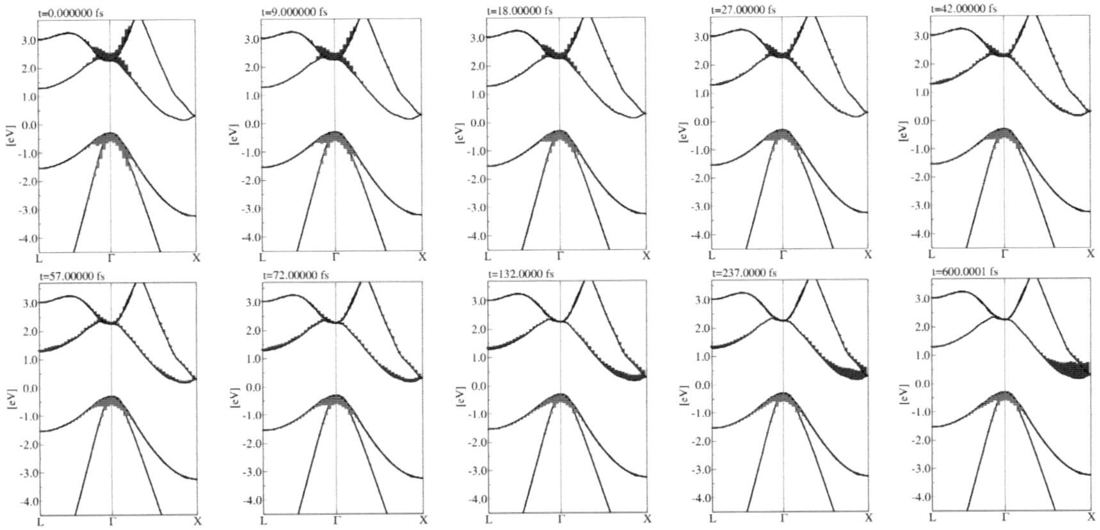

Figure 3. Carriers occupations on the band structure as a function of time

temperature becomes very similar to the case with only the e–p channel active (data not shown).

Finally in Fig. 3 the carriers dynamics is represented on the whole band structure. Here we see from another perspective the difference between the two channels. Holes remain basically close the VBM, just cooling down, while electrons move along the band structure going from Γ to the minimum, nearby X.

In conclusion we have presented a first–principles implementation of carriers dynamics in bulk Silicon. The most critical aspect of the present simulations is the use of a double grid technique for the sampling of the Brilloine zone to guarantee a correct description of the energy transfered from electrons to phonons. The global picture reveal a three systems dynamics: the electrons in the conduction band, the holes in the valence band and the lattice. The three systems exchange energy, electrons an holes vie the e–e scattering, while carriers and phonons via the e–p scattering. Both electrons and holes reach an internal meta–stable state, i.e. they are distributed according to a Fermi distribution, very quickly (approx 0.1 ps). However the complete equilibration, either in between electrons and holes (less then 1 ps) or with the lattice (not yet reached after 1.5 ps), takes longer time. This is, to the best of our knowledge, the first fully ab–initio and parameters free description of carriers dynamics in Silicon.

References

[1] Haug H and Ell C 1992 *Phys. Rev. B* **46**(4) 2126–2132 URL `http://link.aps.org/doi/10.1103/PhysRevB.46.2126`
[2] O'Mara W, Herring R and Hunt L 1990 *Handbook of Semiconductor Silicon Technology* Materials science and process technology series (Noyes Publications) ISBN 9780815512370
[3] Haug H and Jauho A P 2008 *Quantum Kinetics in Transport and Optics of Semiconductors* (Springer)
[4] Sundaram S K and Mazur E 2002 *Nat Mater* **1** 217–224 ISSN 1476-1122 URL `http://dx.doi.org/10.1038/nmat767`
[5] Marini A, Hogan C, Grüning M and Varsano D 2009 *Computer Physics Communications* **180** 1392–1403
[6] Attaccalite C, Grüning M and Marini A 2011 *Phys. Rev. B* **84**(24) 245110 URL `http://link.aps.org/doi/10.1103/PhysRevB.84.245110`
[7] Marini A 2013 *Journal of Physics: Conference Series* **427** 012003 URL `http://stacks.iop.org/1742-6596/427/i=1/a=012003`
[8] Gonze X 1995 *Phys. Rev. A* **52** 1096–1114
[9] Baroni S, de Gironcoli S, Dal Corso A and Giannozzi P 2001 *Rev. Mod. Phys.* **73** 515–562

Journal of Physics: Conference Series **609** (2015) 012007 doi:10.1088/1742-6596/609/1/012007

Magnetic Structure of Mn-doped 6H-SiC

M Al Azri[1], M Elzain[1], K Bouziane[2] and S M Chérif[3]

[1]Department of Physics, College of Science, Sultan Qaboos University, Sultanate of Oman
[2]Pôle Energies renouvelables Environnement et Etudes Pétrolières, Université Internationale de Rabat, Morocco.
[3]LSPM (CNRS-UPR 3407),Université Paris 13, France

E-mail: msk_azri@hotmail.com

Abstract. We investigated the local magnetic properties of Mn-doped 6H-SiC using ab-initio calculations. The calculation addresses various configurations of Mn single impurity and Mn dimers at substitutional/interstitial sites with and without neighboring Si and C vacancies. The calculations showed that a substitutional Mn atom at either Si or C sites possesses a magnetic moment. The Mn atom at Si site possesses larger magnetic moment than Mn atom at C site. Our calculations show that antiferromagnetically coupled pair of Mn atoms at Si sites with neighboring C vacancy is magnetically more stable. The calculation also showed that the interstitial sites with C neighbors are more favorable than those with Si and the magnetic moment for Mn at interstitial sites is less compared to that at substitutional sites. The results are used to understand the experimental data obtained on Mn- 6H-SiC for various Mn concentrations.

1. Introduction

One of the promising classes of materials for application in spintronic are diluted magnetic semiconductors (DMS). Despite much work devoted to the study of DMS, the situation is still controversial from both experimental and theoretical points of view. SiC-based DMSs have attracted considerable interest due to their excellent physical properties due to their extraordinary material properties such as large band gap, high electron saturation velocity, high breakdown field and high thermal conductivity[1]. In addition, SiC is compatible with conventional integrated circuit fabrication. Due to the low solubility of most TM elements in SiC, dopant introduction into SiC is practically limited to ion implantation technique. However, radiation damages such as vacancies, interstitials and extended defects are the undesired effects of ion implantation, which at high ion fluencies can even lead to amorphization of subsurface SiC layers of a few hundred nanometers thickness.

Several reports have been published documenting experimental studies of magnetic properties of SiC with transition metal TM impurities [2-10]. In recent studies, ferromagnetic response was observed in Ni, Mn and Fe implanted 6H-SiC with Tc of 50, 250, and 270 K, for peak concentration of 5% of substitutional TM atoms [2,3]. In addition, 6H-SiC implanted with various doses of Fe showed the samples with the peak Fe concentration below 3% to exhibit DMS behaviour, with the magnetic ordering disappearing above 20 K [4]. In addition, experimental studies showed that Fe implanted 6H-SiC exhibits a high level of substitutional fraction of Fe atoms on Si-sites but without any ferromagnetic response, even at low temperature under 1 T[6]. SiC doped with Mn has become the most actively studied SiC DMS material and a number of experimental studies testify that Mn can be a suitable impurity for achieving a high-temperature FM ordering in the SiC DMS [7-10].

In addition to experimental work, different computational techniques were used recently to study the magnetic properties of SiC DMS. The Si site substitution was found to be more favorable than the

Journal of Physics: Conference Series **609** (2015) 012007 doi:10.1088/1742-6596/609/1/012007

C site for TM substitution in 3C-SiC [11,12]. Los et al [13] studied the magnetic properties of Fe-doped 4H-SiC using FLAPW method, and they found that Fe can be in either a magnetic or nonmagnetic state and this depends on the TM atom environment in the host matrix. Zhou et al[14] studied the structural and magnetic properties of 3d transition metal doped 3C-SiC using ab initio calculations. They found that the structure of substitutional TM_{Si} is the most stable one for early transition metals like Ti, V, Cr, and Mn, while the clustering of TM_{Si}-TM_I dimers formed by the neighboring substitutional TM_{Si} and interstitial TM_I is energetically favored for late transition metals such as Co, Ni, and Cu. Most of theoretical work done so far concentrated on 3C-SiC with few results relating to TM doped 6H-SiC. We have studied recently the influence of the implantation fluence on the generation of defects and on the Mn fraction at substitutional Si site (Mn_{Si}) and its effect on the local magnetic moment of Mn[15]. It was shown that as Mn fluence increases, Mn in substitutional sites increases up to certain fluence. Further increase in the fluence leads to an increase in the accumulated damage that suppresses the magnetization. Reduction of magnetization of all samples was observed with heat treatment [16]. It is clear that the defects induced by the implantation play the major role in the magnetic properties of the as-implanted and annealed samples. To have a clear picture about the correlation between structural and magnetic properties, we investigated the structural and magnetic properties of Mn-doped 6H-SiC using ab-initio calculations. In this paper, we report the results of the magnetic properties of Mn doped 6H-SiC. Various configurations of Mn sites have been considered. Magnetic coupling between the Mn atom with and without neighboring vacancy, to Mn at substitutional and/or interstitial sites were explored.

2. Methodology

We consider the 6H polytype of SiC, which has a hexagonal structure with 12 bilayers in the unit cell. The layers are stacked in hexagonal (*h*) and cubic (*k*) sequences as $hk_1k_2hk_1k_2$. Tetrahedral configurations and the number of nearest neighbors are maintained within all stacking.

In this work, we used the FP-LAPW method as implemented in WIEN2k package [17] to study the magnetic properties of Mn-doped 6H-SiC. The embedded impurities in SiC are modeled by a 2×2×1 supercell. The muffin-tin radii (RMT) have been chosen as 1.66 au for Si and C and 1.77 au for Mn. An optimum number of k-points was reached by testing the convergence of the total energy and the electric field gradient for the basic unit cell of 6H-SiC with two Mn as substitutional atoms. The supercell calculations were performed with an energy cut-off such that RMTkmax= 6.0 and a k sampling with a 8×8×2 Monkhost-Pack mesh. The Kohn-Sham equations were solved using the Perdew-Burke-Ernerhof GGA approximation [18]. The self-consistent cycles were stopped when the set convergence criteria in the total energy and the charge density are reached. Various configurations were studied including a single Mn impurity and Mn dimers at substitutional/interstitial sites with/without neighboring vacancies.

3. Results and Discussions:

The main defects introduced by Mn implantation into 6H-SiC are vacancy type defects and interstitials [19]. Consequently, we studied two cases of substitution: (i) Mn at Si/C site without neighboring vacancy, and (ii) Mn at Si/C site with neighboring C/Si vacancy. In addition, Mn at interstitial positions where it can be surrounded by four C/Si atoms with and without C/Si vacancy was also studied. The calculated magnetic moments for each configuration are presented in table 1. The first column represents the substitutional site and the second column is the vacancy site. Here MMI stands for the moment within the muffin tin sphere.

It is clear from the table that the magnetic moment depends on whether the dopant substitutional Mn atom is at C or Si site. In absence of vacancies, the magnetic moment of Mn at C site is smaller as compared to that at Si site in agreement with reported results [20]. In comparison to the corresponding results for Fe in 6H-SiC, it was found that Fe at the substitutional sites without neighbouring vacancy are nonmagnetic [21].

Table 1. The dopant site, vacancy site and magnetic moments (μ_B) in 6H-SiC doped with Mn atom. MMI is the magnetic moment per 3d-atom. Mn_I(C) refers to Mn at interstitial site with C neighbors and Mn_I(Si) refers to Mn at interstitial site with Si neighbours.

Dopant site	Vacancy site	MMI(μB)
	Substitutional site	
Si	-	2.31
C	-	0.68
Si	C	2.68
C	Si	1.44
	Interstitial site	
Mn_I (C)	-	1.52
Mn_I(Si)	-	0.48
Mn_I (C)	C	0.45
Mn_I(Si)	Si	0.60

In order to simulate the introduction of defects by ion implantation and to be able to compare the energies, vacancies are introduced at neighbouring Si/C sites. In this case it is possible to compare the energies of supercells. We found that Mn at the Si site with a neighbouring C vacancy is far more stable than Mn at the C site with Si. This may be justified by the large differences in atomic radii of Mn and C, on one hand, and their similarities for Mn and Si, on the other hand. The Si site substitution preference is in agreement with previous studies[11-13]. The magnetic moment of Mn increases with introduction of vacancies.

Table 1 shows that Mn interstitial sites has smaller magnetic moment compared to the substitutional sites, which explains the decrease of the magnetic moment by increasing the interstitial defects. We also found that the Mn at interstitial sites with C neighbors are more favorable than those with Si neighbors with an energy difference of about 2.0 eV. On the other hand introduction of vacancies at neighboring sites leads to further reduction in magnetic moment in contrast to the case of substitutional sites. It should be pointed out that these results are consistent with the experimental results for the as-implanted samples. Indeed, we found that the sample with lower fluence, even though most Mn atoms occupy C substitutional site, the relatively low Mn concentration (0.7%) leads to a small magnetic moment. As the fluence increases, the ratio of Mn atoms occupying substitutional sites increases, corresponding to higher Mn concentration (~2%), and the Mn substitute Si site giving relatively higher magnetic moment.

We have also studied the behaviour of Mn dimers in 6H-SiC. We found that antiferromagnetically coupled dimers are more stable than those ferromagnetically coupled. The details of these results will be published elsewhere.

4. Conclusion

The FP-LAPW method is used to study the local magnetic properties of Mn doped 6H-SiC. Substitutional Mn atom at Si atom gives higher magnetic moment compared to the case when Mn substitutes a C atom and at interstitial sites. We found that magnetic moment increases when there is a neighboring C/Si vacancy of Mn at Si site and Mn at C site respectively. The calculations also showed that the AFM coupled Mn atoms at substitutional sites are more stable than those FM coupled.

References:

[1] Choyke W J and Devaty R P 1997 *Diamond and Related Materials* **6** (10) 1243-48.

[2] Theodoropoulou N, Hebard A F, Chu S N G, Overberg M E, Abernathy C R, Pearton S J, Wilson R G and Zavada J M 2001, *Electrochemical and Solid-State Letters* **4** (12) G119-G121.

[3] Theodoropoulou N, Hebard A F, Chu S N G, Overberg M E , Abernathy C R, Pearton S J, Wilson R G, Zavada J M and Park Y D 2002, *J. Vac. Sci.Technol.* A*: Vac., Surfaces, and Films* **20** (3) 579-582.

[4] Stromberg F, Keune W, Chen X, Bedanta S, Reuther H and Mücklich A 2006 *J. Phys: Condens. Matter.* **18** (43) 9881.

[5] Lin S, Chen Z, Liang P, Jiang D and Xie H 2010 *Chem .Phys. Lett.* **496** (1–3) 56-58.

[6] Dupeyrat C, Declémy A, Drouet M, Eyidi D, Thomé L, Debelle A,Viret M and Ott F 2009 *Physica B: Condens. Matter.* **404** (23-24) 4731-34.

[7] Bouziane K, Mamor M, Elzain M, Djemia P and Chérif S M 2008 *Phys. Rev. B* **78** (19) 195305.

[8] Wang W, Takano F, Ofuchi H and Akinaga H 2008 *New J. Phy.* **10** (5) 055006.

[9] Song B, Bao H, Li H, Lei M, Jikang J, Han J, Zhang X, Songhe M, Wang W and Chen X 2009 *App. Phys. Lett.* **94** (10) 102503-08.

[10] Ma S B, Sun Y P, Zhao B C, Tong P, Zhu X B and Song W H 2007 *Physica B: Condens. Matter.* **94** (1) 122-126.

[11] Gubanov V A, Boekema C and Fong C Y, 2001 *App.Phys. Lett.* **78** (2) 216-218.

[12] Miao M S and Lambrecht W R L, 2003 *Phys. Rev.* B **68** (12) 125204.

[13] Los A V, Timoshevskii A, Los V F and Kalkuta S A 2007 *Phys. Rev.* B **76** (16) 165204.

[14] Zhou P L, Zheng S K, Ma Lei, Jingfang, He Yan Tian and Shi Ruian 2014 *Solid State Communications* **196** 28-31.

[15] Bouziane K, Al Azri M, Elzain M, Chérif S M, Mamor M, Declémy A and Thomé L 2015 *J. Alloys and Compounds* **632** 760–765.

[16] Al Azri M, Elzain M, Bouziane K, Chérif M, Declémy A,Thomé L, and Viret M 2014 *IEEE Trans. Magn.* **50**(11).

[17] Blaha P, Schwarz K, Madsen G K H, Kvasnicka D and Luitz J, 2001 *WIEN2K, An Augmented Plane Wave Plus Local Orbitals Program for Calculating Crystal Properties* (Karlheinz Schwarz,Technical Universitat Wien, Austria).

[18] Perdew J P,Burke K and Ernzerhof M 1996 *Phys. Rev. Lett.* **77** (18) 3865-68.

[19] Chadwick A V, Terenzi M (Eds.) 1986 *Defects in Solids, ModernTechniques, NATO ASI Series, Series B: Physics* **147**.

[20] Shaposhnikov V L and Sobolev N A 2004 *J. Phy.: Condens. Matter* **16** (10) 1761.

[21] Elzain M, Al-Harthi S H , Gismelseed A, Al-Rawas A, Yousif A, Widatallah H, Al-Barwani M 2014 *Hyperfine Interactions* **226** 281-287.

Journal of Physics: Conference Series **609** (2015) 012008 doi:10.1088/1742-6596/609/1/012008

Strain effects on effective masses for MoS_2 monolayers

E. R. Ortiz[1], Blanca Biel[1], Luca Donetti[1], Andrés Godoy[1], Francisco Gámiz[1]
[1]Dpto. Electrónica y Tecnología de Computadores, Facultad de Ciencias, Campus de Fuente Nueva, & CITIC, Campos de Aynadamar, 18071 Universidad de Granada, Spain

E-mail: biel@ugr.es

Abstract: We have studied the effects of both intrinsic and external factors, such as strain, substrate and electric field, over a MoS_2 monolayer, via DFT simulations. From these simulations we computed the system's bandstructure and evaluate the bandgap values, extracting the effective masses as well. Our calculations show a transition between direct and indirect gaps when strain is applied, and also that the bandstructure is barely affected by the presence of a substrate or an external electric field with values up to 1 V/Å .

1. Introduction

In recent years, two-dimensional (2D) materials, such as graphene, have drawn great attention for their possible applications in nanoelectronics. As the name implies, these materials are formed by a single atomic or molecular layer, with different layers stacking together under the effect of the so called van der Waals forces, which can be used to create complex heterostructures [1-3] while keeping the side effects of coupling different materials at a minimum.

One of those materials is molybdenum bisulfide (MoS_2), where the main monolayer is formed by three sublayers, as shown in Fig. 1. MoS_2, unlike graphene, behaves as a semiconductor with indirect gap (bulk) or direct gap at the K point (monolayer) [4-6], which may prove useful for the design of nanoelectronic devices.

Therefore, tuning these properties is essential if MoS_2 is to be implemented in future nanoelectronic devices. In this work, we study the effect on the bandstructure and effective masses of both structural changes, via straining, or external, in the form of a perpendicular electric field or a silicon oxide substrate.

2. Methodology

All the DFT calculations have been performed using SIESTA [7], within the Generalized Gradient Approximation (GGA) and the pseudopotential approaches.

We have used the SIMPLEX algorithm [8] to optimize the DZP basis set for monolayer MoS_2 and SiO_2, in their experimental lattice constants of 3.160 Å and 4.914 Å, respectively. This lattice constant corresponds to the silanol reconstruction of crystalline α-quartz, which has been widely used in previous works [9, 10].

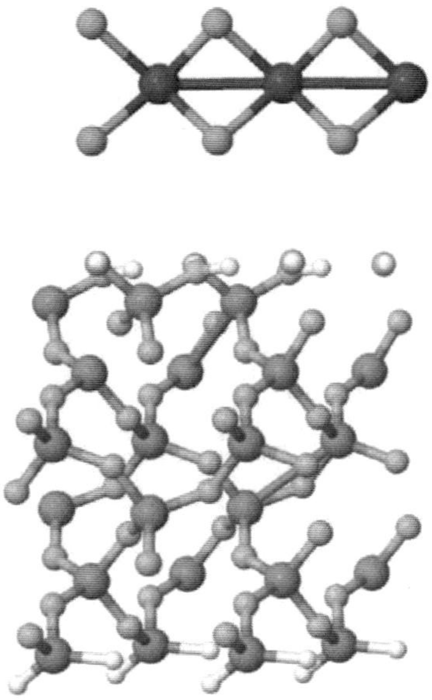

Figure 1. Top: side view of the MoS_2 monolayer. Sulfur is marked as orange (light) atoms, while molybdenum is represented as a green (dark) atom. 3 unit cells are represented.
Bottom: silanol substrate. Silicon atoms are represented on blue (dark) and oxygen as green (light). The white atoms on the surfaces represent the hydrogen atoms.

To accomplish with the 3D periodicity in SIESTA, a 2D monolayer requires a manual separation on the perpendicular z axis, 20 Å in this case, to avoid interaction between different layers. This also affects the Brillouin zone sampling, where we used a Monkhorst-Pack scheme of 9x9x1 points. We relaxed the structure until atomic forces were bellow 0.01 eV/ Å. The obtained bandgap value for unstrained MoS_2 is 1.81 eV, close (albet accidentally [11-13]) to the experimental 1.8 eV [4-6].

After the first monolayer simulation, we performed several strain tests using strain values of 3.66% and -3%, applied on both x and y axes. The first value corresponds to the expected strain caused by an α-quartz substrate as described above. These simulations were later repeated, adding an external electric field of 0.25, 0.5 and 1 eV/ Å and a silanol substrate.

In all these cases we also calculated the effective masses on both valence and conduction bands, using a parabolic approach to obtain the band curvature at the special points in the reciprocal space.

3. Results

3.1. Strain effects on the bandstructure.

In previous sections, we introduced bulk MoS_2 as an indirect gap semiconductor, while unstrained, monolayer MoS_2 behaves as a direct gap semiconductor with a 1.8 eV gap. However, once the biaxial strain has been applied, the gap becomes indirect once again, as shown in Fig. 2. Fig. 2 contains four calculated bandstructures along the Γ-M-K-Γ path of the Brillouin zone, for unstrained MoS_2, compressive and tensile strain. For compressive strain, our simulations show a displacement of the conduction band minimum (CBM) from the original K point to a midpoint between the K and Γ

Journal of Physics: Conference Series **609** (2015) 012008 doi:10.1088/1742-6596/609/1/012008

points, resulting on an indirect gap, as already reported on previous works[14, 15]. For tensile strain, in both cases the valence band maximum (VBM) switches from the K point to the $\bar{\Gamma}$ point, leading to another indirect bandgap. For all the studied strain values, the value of the direct bandgap at the K point decreases, from a maximum of 2.16 eV while compressed to a minimum of 1.44 while tensed at the maximum strain considered on this work. Fig. 3 shows a more detailed description, including other possible transitions.

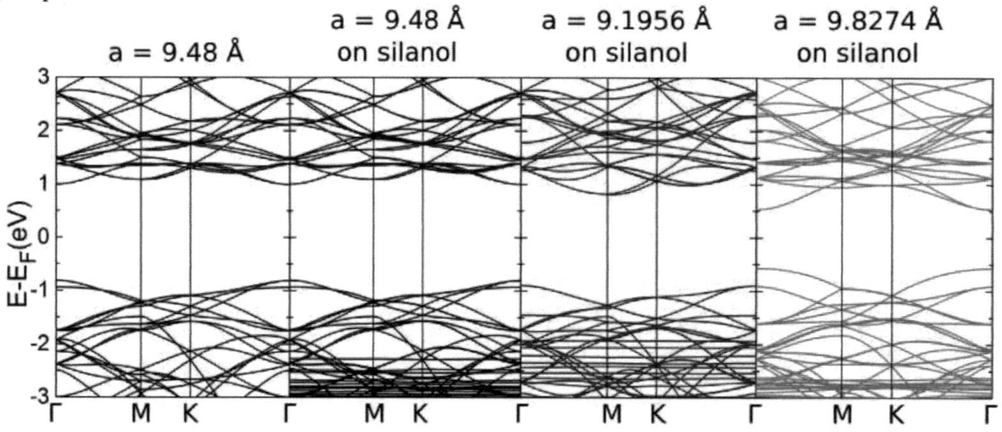

Figure 2. Monolayer MoS_2 bandstructure for different strains. From left to right: isolated, unstrained monolayer; unstrained, tensely and compressively strained on a silanol substrate. To match sizes with the substrate a 3x3 supercell was used to recreate the MoS_2 layer.

3.2. Effective masses.

To calculate the effective masses, we fitted the extremes of the obtained bandstructures at the relevant symmetry points of the Brillouin zone. Our results are summarized in Fig. 3, which contains the different values for electron and holes mass in all of those points, for all the different strains. In particular, the results obtained for hole mass in the unstrained sample are in good agreement with the literature, with m^*/m_0=0.67 for the K point and m^*/m_0=3.34 for the $\bar{\Gamma}$ point, versus the m^*/m_0=0.64 and m^*/m_0=3.52 from previous works[14].

3.3. SiO_2 substrate and external electric field

In order to study the effects of external factors, we set an external electric field on the perpendicular direction of the 2D layer, with values ranging from 0.25 V/Å to 1V/Å, for all the different afore mentioned strain values. For all these cases, the field does not induce appreciable modifications on the bandstructure.

As for the silanol substrate, our results match once again those proposed in the literature [16], showing that the presence of a substrate does not affect the bandstructure enough to cause a change the effective masses, which remain the same as in the previous sections but for the presence of the oxide bands for energies far form the Fermi level. In Figure 2, the bandstructures for different strain values are shown, with and without the presence of the substrate, for comparison. Other than the change caused from the increased supercell, the states created by silanol remain away from the bandgap, causing no remarkable changes.

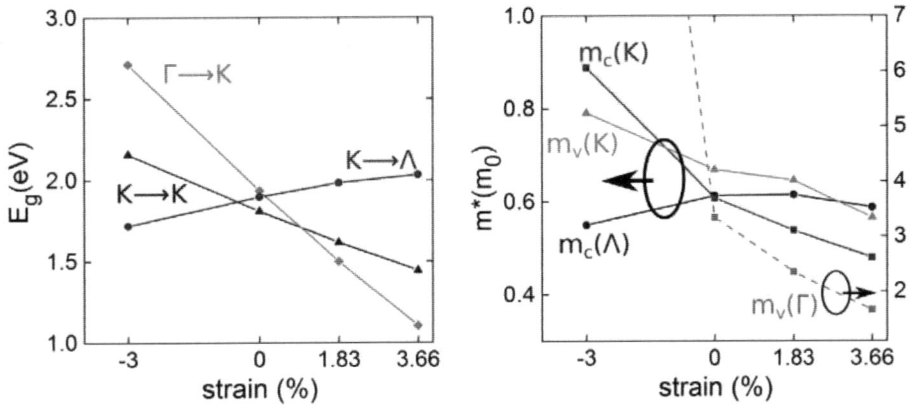

Figure 3. Left panel: bandgap values for the transitions between VBM and CBM at the relevant points of the Brilloun zone. Right panel: effective masses for holes and electrons on each relevant point of the Brillouin zone for the studied strain values.

4. Conclusions

We calculated the bandstructure for MoS_2 under different strain effects, external electric fields and on a silanol substrate. From such bandstructures the effective masses were calculated for the relevant symmetry points of the Brillouin zone. Strain simulations show a transition from direct to indirect gap, while neither the electric field nor the silanol substrate barely affect the bandstructure.

5. Acknowledgements

B. Biel thankfully acknowledges funding by the RYC-2012-10381 contract from Ramón y Cajal Program (MINECO) and computer facilities provided by the RES-BSC and the CSIRC-UGR supercomputing centers. A. Godoy acknowledges funding by the MINECO contract FIS 2011-26005.

6. References

[1] A. K. Geim and I. V. Grigorieva, Nature 499 (2013), 419
[2] A. Hsu et al., Proceedings of the IEEE 101 (2013), 1638
[3] G. Fiori et al. Nature Nanotechnol. 9 (2014), 768
[4] K. F. Mak, et al., Phys. Rev. Lett. 105 (2010), 136805
[5] A. Splendiani et al., Nano Lett. 10 (2010), 1271
[6] A. Kumar and P. K. Ahluwalia, Eur. Phys. J. B 85 (2012), 186
[7] J.M. Soler et al., J. Phys.: Condens. Matter 14 (2002), 2745
[8] W. Press et al., Numerical Recipes, Cambridge U. Press, (1989)
[9] M.H. Evans et al. Phys. Rev. Lett. 95 (2005), 106802
[10] T. Hara et al., J. Phys. Conf. Ser. 109 (2008), 012012
[11] K. Dolui et al. Phys. Rev. B 85 (2012) 205302
[12] A. Ramasubramaniam, Phys. Rev. B 86 (2012) 115409
[13] T. Cheiwchanchamnangij and W. R. L. Lambrecht, Phys. Rev. B 85 (2012) 205302
[14] W. S. Yun et al., Phys. Rev. B 85 (2012), 033305
[15] N. Harada et al., J. Appl. Phys. 115 (2014), 034505
[16] K. Dolui et al., Phys. Rev. B 87 (2013), 16540

AUTHOR INDEX